IFIP Advances in Information and Communication Technology　486

Editor-in-Chief

Kai Rannenberg, Goethe University Frankfurt, Germany

IFIP – The International Federation for Information Processing

IFIP was founded in 1960 under the auspices of UNESCO, following the first World Computer Congress held in Paris the previous year. A federation for societies working in information processing, IFIP's aim is two-fold: to support information processing in the countries of its members and to encourage technology transfer to developing nations. As its mission statement clearly states:

IFIP is the global non-profit federation of societies of ICT professionals that aims at achieving a worldwide professional and socially responsible development and application of information and communication technologies.

IFIP is a non-profit-making organization, run almost solely by 2500 volunteers. It operates through a number of technical committees and working groups, which organize events and publications. IFIP's events range from large international open conferences to working conferences and local seminars.

The flagship event is the IFIP World Computer Congress, at which both invited and contributed papers are presented. Contributed papers are rigorously refereed and the rejection rate is high.

As with the Congress, participation in the open conferences is open to all and papers may be invited or submitted. Again, submitted papers are stringently refereed.

The working conferences are structured differently. They are usually run by a working group and attendance is generally smaller and occasionally by invitation only. Their purpose is to create an atmosphere conducive to innovation and development. Refereeing is also rigorous and papers are subjected to extensive group discussion.

Publications arising from IFIP events vary. The papers presented at the IFIP World Computer Congress and at open conferences are published as conference proceedings, while the results of the working conferences are often published as collections of selected and edited papers.

IFIP distinguishes three types of institutional membership: Country Representative Members, Members at Large, and Associate Members. The type of organization that can apply for membership is a wide variety and includes national or international societies of individual computer scientists/ICT professionals, associations or federations of such societies, government institutions/government related organizations, national or international research institutes or consortia, universities, academies of sciences, companies, national or international associations or federations of companies.

More information about this series at http://www.springer.com/series/6102

Zhongzhi Shi · Sunil Vadera
Gang Li (Eds.)

Intelligent Information Processing VIII

9th IFIP TC 12 International Conference, IIP 2016
Melbourne, VIC, Australia, November 18–21, 2016
Proceedings

 Springer

Editors
Zhongzhi Shi
Chinese Academy of Sciences
Beijing
China

Gang Li
Deakin University
Burwood, VIC
Australia

Sunil Vadera
University of Salford
Salford
UK

ISSN 1868-4238   ISSN 1868-422X   (electronic)
IFIP Advances in Information and Communication Technology
ISBN 978-3-319-83930-1   ISBN 978-3-319-48390-0   (eBook)
DOI 10.1007/978-3-319-48390-0

Printed on acid-free paper

This Springer imprint is published by Springer Nature
The registered company is Springer International Publishing AG
The registered company address is: Gewerbestrasse 11, 6330 Cham, Switzerland

Preface

This volume comprises the 9th IFIP International Conference on Intelligent Information Processing. As the world proceeds quickly into the Information Age, it encounters both successes and challenges, and it is well recognized that intelligent information processing provides the key to the Information Age and to mastering many of these challenges. Intelligent information processing supports the most advanced productive tools that are said to be able to change human life and the world itself. However, the path is never a straight one and every new technology brings with it a spate of new research problems to be tackled by researchers; as a result we are not running out of topics; rather the demand is ever increasing. This conference provides a forum for engineers and scientists in academia, university and industry to present their latest research findings in all aspects of intelligent information processing.

We received more than 40 papers, of which 24 papers are included in this program as regular papers and 3 as short papers. We are grateful for the dedicated work of both the authors and the referees, and we hope these proceedings will continue to bear fruit over the years to come. All papers submitted were reviewed by two referees.

A conference such as this cannot succeed without the help from many individuals who contributed their valuable time and expertise. We want to express our sincere gratitude to the Program Committee members and referees, who invested many hours for reviews and deliberations. They provided detailed and constructive review reports that significantly improved the papers included in the program.

We are very grateful the sponsorship of the following organizations: IFIP TC12, Deakin University, and Institute of Computing Technology, Chinese Academy of Sciences. Thanks to Gang Ma for carefully checking the proceedings.

Finally, we hope you find this volume inspiring and informative.

August 2016

Zhongzhi Shi
Sunil Vadera
Gang Li

Organization

General Chair

E. Chang (Australia)

Program Chairs

Z. Shi (China)
S. Vadera (UK)
G. Li (Australia)

Program Committee

A. Aamodt (Norway)
B. An (Singapore)
A. Bernardi (Germany)
L. Cao (Australia)
E. Chang (Australia)
L. Chang (China)
E. Chen (China)
H. Chen (China)
Z. Cui (China)
T. Dillon (Australia)
S. Ding (China)
Y. Ding (USA)
Q. Dou (China)
E. Ehlers (South Africa)
P. Estraillier (France)
U. Furbach (Germany)
Y. Gao (China)
T. Hong (Taiwan)
Q. He (China)
T. Honkela (Finland)
Z. Huang
  (The Netherlands)
G. Kayakutlu (Turkey)
D. Leake (USA)
G. Li (Australia)
J. Li (Australia)
Q. Li (China)
W. Li (Australia)

X. Li (Singapore)
J. Liang (China)
Y. Liang (China)
H. Leung (HK)
P. Luo (China)
H. Ma (China)
S. Ma (China)
W. Mao (China)
X. Mao (China)
Z. Meng (China)
E. Mercier-Laurent
  (France)
D. Miao (China)
S. Nefti-Meziani (UK)
W. Niu (China)
M. Owoc (Poland)
G. Pan (China)
H. Peng (China)
G. Qi (China)
A. Rafea (Egypt)
ZP. Shi (China)
K. Shimohara (Japan)
A. Skowron (Poland)
M. Stumptner (Australia)
K. Su (China)
D. Tian (China)
I. Timm (Germany)
H. Wei (China)

P. Wang (USA)
G. Wang (China)
S. Tsumoto (Japan)
J. Weng (USA)
Z. Wu (China)
S. Vadera (UK)
Y. Xu (Australia)
Y. Xu (China)
H. Xiong (USA)
X. Yang (China)
Y. Yang (Australia)
Y. Yao (Canada)
W. Yeap (New Zealand)
J. Yu (China)
B. Zhang (China)
C. Zhang (China)
L. Zhang (China)
M. Zhang (Australia)
S. Zhang (China)
Z. Zhang (China)
Y. Zhao (Australia)
Z. Zheng (China)
C. Zhou (China)
J. Zhou (China)
Z.-H. Zhou (China)
J. Zhu (China)
F. Zhuang (China)
J. Zucker (France)

Keynote and Invited Presentations
(Abstracts)

Automated Reasoning and Cognitive Computing

Ulrich Furbach

University of Koblenz, Mainz, Germany
uli@furbach.de

Abstract. This talk discusses the use of first order automated reasoning in question answering and cognitive computing. The history of automated reasoning systems and the state of the art are sketched. In a first part of the talk the natural language question answering project LogAnswer is briefly depicted and the challenges faced therein are addressed. This includes a treatment of query relaxation, web-services, large knowledge bases and co-operative answering. In a second part a bridge to human reasoning as it is investigated in cognitive psychology is constructed; some examples from human reasoning are discussed together with possible logical models. Finally the topic of benchmark problems in commonsense reasoning is presented together with our appoach.

Keywords: Automated reasoning · Cognitive computing · Question answering · Cognitive science · Commonsense reasoning

An Elastic, On-demand, Data Supply Chain for Human Centred Information Dominance

Elizabeth Chang

The University of New South Wales, Canberra, Australia
elizabeth.chang@unsw.edu.au

Abstract. We consider different instances of this broad framework, which can roughly be classified into two cases. In one instance, the system is assumed to be a black box, whose inner working is not known, but whose states can be (partially) observed during a run of the system. In the second instance, one has (partial) knowledge about the inner working of the system, which provides information on which runs of the system are possible. In this talk, we will review some of our recent research that investigates different instances of this general framework of ontology-based monitoring of dynamic systems. Getting the right data from any data sources, in any formats, with different sizes and have different multitudes of complexity, in real time to the right person at the right time and in a form which they can rapidly assimilate and use is the concept of Elastic On-demand Data Supply Chain. Finding out what data is needed from which system, where and why is it needed, how is the data searched, extracted, aggregated represented and how should it be presented visually so that the user can use and operate the information without much training is applying a human centred approach to on-demand data supply chain. Information Dominance represents how by using guided analytics and self-service on the data, human cognitive information capabilities including optimization of systems and resources for decision making in the dynamic and complex environment are built. In this presentation, I explain these concepts and demonstrate how the effectiveness and efficiency of the above integrated approach is validated by providing both theoretical concept proofing with stratification, target sets, reachability, incremental enlargement principle and practical concept proofing through implementation of the Faceplate. The project is funded by Australian Department of Defence.

Why Is My Entity Typical or Special? Approaches for Inlying and Outlying Aspects Mining

James Bailey

Department of Computing and Information Systems,
The University of Melbourne, Parkville, Australia
baileyj@unimelb.edu.au

Abstract. When investigating an individual entity, we may wish to identify aspects in which it is usual or unusual compared to other entities. We refer to this as the inlying/outlying aspects mining problem and it is important for comparative analysis and answering questions such as "How is this entity special?" or "How does it coincide or differ from other entities?" Such information could be useful in a disease diagnosis setting (where the individual is a patient) or in an educational setting (where the individual is a student). We examine possible algorithmic approaches to this task and investigate the scalability and effectiveness of these different approaches.

Advanced Reasoning Services
for Description Logic Ontologies

Kewen Wang

School of Information Technology, Griffith University, Nathan, Australia
k.wang@griffith.edu.au

Abstract. Ontology-like knowledge bases (KBs) have become a promising modeling tool in a wide variety of applications such as intelligent Web search, question understanding, in-context advertising, social media mining, and biomedicine. Such KBs are distinct from traditional KBs in that they are based on *ontologies* (as schemas) that assist in organization and access of information on the Web and from other sources. However, practical ontology-like KBs are usually associated with data of large volume, dynamic with content, and updated rapidly. Efficient systems have been developed for standard reasoning and query answering for OWL/Description Logic (DL) ontologies. In recent years, the issue of facilitating advanced reasoning services is receiving extensive attention in the research community. In this talk, we will discuss recent research results and challenges of three important reasoning tasks of ontologies including ontology change, query explanation and rule-based reasoning for OWL/DL ontologies.

Brain-Like Computing

Zhongzhi Shi

Key Laboratory of Intelligent Information Processing, Institute of Computing Technology, Chinese Academy of Sciences, Beijing, 100190, China
shizz@ics.ict.ac.cn

Abstract. Human-level artificial intelligence, which makes machines with intelligent behavior of the human brain, is the most challenging major scientific issues of this century, but also is the current hot topics in academic and industry area. Brain-like computing has become the leading edge technology in twenty-first Century, many countries have started the brain science and cognitive computing projects. Intelligence science has brought a number of inspiration to the machine intelligence, and promote the research on brain science, cognitive science, intelligent computing technology and intelligent robot. In this talk, I will focus on the research progress and development trend of cognitive models, brain-machine collaboration, and brain-like intelligence.

Brain-like intelligence is a new trend of artificial intelligence that aims at human-level artificial intelligence through modeling the cognitive brain and obtaining inspiration from it to power new generation intelligent systems. In recent years, the upsurges of brain science and intelligent technology research have been developed in worldwide.

Acknowledgements. This work is supported by the National Program on Key Basic Research Project (973) (No. 2013CB329502).

Contents

Social Computing

Semantic Web and Text Processing

Image Understanding

Machine Learning

An Attribute-Value Block Based Method of Acquiring Minimum Rule Sets: A Granulation Method to Construct Classifier

Zuqiang Meng[(⊠)] and Qiuling Gan

College of Computer, Electronics and Information, Guangxi University,
Nanning 530004, Guangxi, China
zqmeng@126.com

Abstract. Decision rule acquisition is one of the important topics in rough set theory and is drawing more and more attention. In this paper, decision logic language and attribute-value block technique are introduced first. And then realization methods of rule reduction and rule set minimum are relatively systematically studied by using attribute-value block technique, and as a result effective algorithms of reducing decision rules and minimizing rule sets are proposed, which, together with related attribute reduction algorithm, constitute an effective granulation method to acquire minimum rule sets, which is a kind classifier and can be used for class prediction. At last, related experiments are conducted to demonstrate that the proposed methods are effective and feasible.

Keywords: Rule acquisition · Attribute-value blocks · Decision rule set · Classifier

1 Introduction

Rough set theory [1], as a powerful mathematical tool to deal with insufficient, incomplete or vague information, has been widely used in many fields. In rough set theory, the study of attribute reduction seems to attract more attention than that of rule acquisition. But in recent years there have been more and more studies involving the decision rule acquisition. Papers [2, 3] gave discernibility matrix or the discernibility function-based methods to acquire decision rules. These methods are able to acquire all minimum rule sets for a given decision system theoretically, but they usually would pay both huge time cost and huge space cost, which extremely narrow their applications in real life. In addition, paper [4] discussed the problem of producing a set of certain and possible rules from incomplete data sets based on rough sets and gave corresponding rule learning algorithm. Paper [5] discussed optimal certain rules and optimal association rules, and proposed two quantitative measures, random certainty factor and random coverage factor, to explain relationships between the condition and decision parts of a rule in incomplete decision systems. Paper [6] also discussed the rule acquisition in incomplete decision contexts. This paper presented the notion of an approximate decision rule, and then proposed an approach for extracting non-redundant approximate decision rules from an incomplete decision context. But the proposed

Z. Shi et al. (Eds.): IIP 2016, IFIP AICT 486, pp. 3–11, 2016.
DOI: 10.1007/978-3-319-48390-0_1

method is also based on discernibility matrix and discernibility function, which determines that it is relatively difficult to acquire decision rules from large data sets.

Attribute-value block technique is an important tool to analyze data sets [7, 8]. Actually, it is a granulation method to deal with data. Our paper will use the attribute-value block technique and other related techniques to systematically study realization methods of rule reduction and rule set minimum, and propose effective algorithms of reducing decision rules and minimizing decision rule sets. These algorithms, together with related attribute reduction algorithm, constitute an effective solution to the acquisition of minimum rule sets, which is a kind classifier and can be used for class prediction.

The rest of the paper is organized as follows. In Sect. 2, we review some basic notions linked to decision systems. Section 3 introduces the concept of minimum rule sets. Section 4 gives specific algorithms for rule reduction and rule set minimum based on attribute-value blocks. In Sect. 5, some experiments are conducted to verify the effectiveness of the proposed methods. Section 6 concludes this paper.

2 Preliminaries

In this section, we first review some basic notions, such as attribute-value blocks, decision rule sets, which are prepared for acquiring minimum rule sets in next sections.

2.1 Decision Systems and Relative Reducts

A decision system (DS) can be expressed as the following 4-tuple: $DS = (U, A = C \cup D, V = \bigcup_{a \in A} V_a, \{f_a\})$, where U is a finite nonempty set of objects; C and D are condition attribute set and decision attribute set, respectively, and $C \cap D = \emptyset$; V_a is a value domain of attribute a; $f_a : U \to V$ is an information function from U to V, which maps an object in U to a value in V_a.

For simplicity, $(U, A = C \cup D, V = \bigcup_{a \in A} V_a, \{f_a\})$ is expressed as $(U, C \cup D)$ if V and f_a are understood. Without loss of generality, we suppose D is supposed to be composed of only one attribute.

For any $B \subseteq C$, let $U/B = \{[x]_B \mid x \in U\}$, where $[x]_B = \{y \in U \mid f_a(y) = f_a(x)$ for any $a \in B\}$, which is known as equivalence class. For any subset $X \subseteq U$, the lower approximation $\underline{B}X$ and the upper approximation $\overline{B}X$ of X with respect to B are defined by: $\underline{B}X = \{x \in U \mid [x]_B \subseteq X\}$, $\overline{B}X = \{x \in U \mid [x]_B \cap X \neq \phi\}$. And then the concepts of positive region $POS_B(X)$, boundary region $BND_B(X)$ and negative region $NEG_B(X)$ of X are defined as: $POS_B(X) = \underline{B}X$, $BND_B(X) = \overline{B}X - \underline{B}X$, $NEG_B(X) = U - \overline{B}X$.

Suppose that $U/D = \{[x]_D \mid x \in U\} = \{D_1, D_2, \ldots, D_m\}$, where $m = |U/D|$, D_i is a decision class, $i \in \{1, 2, \ldots, m\}$. Then for any $B \subseteq C$, the concepts of positive region $POS_B(D)$, boundary region $BND_B(D)$ and negative region $NEG_B(D)$ of a decision system $(U, C \cup D)$ can be defined as follows:

$$POS_B(D) = POS_B(D_1) \cup POS_B(D_2) \cup \ldots \cup POS_B(D_m),$$
$$BND_B(D) = BND_B(D_1) \cup BND_B(D_2) \cup \ldots \cup BND_B(D_m),$$
$$NEG_B(D) = U - POS_B(D) \cup BND_B(D).$$

With the positive region, the concept of reducts can be defined as follows: given a decision system $(U, C \cup D)$ and $B \subseteq C$, B is a **relative reduct** of C with respect to D if the following conditions are satisfied: (1) $POS_B(D) = POS_C(D)$, and (2) for any $a \in B, POS_{B-\{a\}}(D) \neq POS_B(D)$.

2.2   Decision Logic and Attribute-Value Blocks

Decision rules are in fact related formulae in decision logic. In rough set theory, a decision logic language depends on a specific information system, while a decision system $(U, C \cup D)$ can be regarded as being composed of two information systems: (U, C) and (U, D). Therefore, there are two corresponding decision logic languages, while attribute-value blocks just act as a bridge between the two languages. For the sake of simplicity, let $IS(B) = (U, B, V = \bigcup_{a \in A} V_a, \{f_a\})$ is an information system with respect to B, where $B \subseteq C$ or $B \subseteq D$. Then a decision logic language $DL(B)$ is defined as a system being composed of the following formulae [3]:

(1) (a, v) is an atomic formula, where $a \in B, v \in V_a$;
(2) an atomic formula is a formula in $DL(B)$;
(3) if φ is a formula, then $\sim \varphi$ is also a formula in $DL(B)$;
(4) if both φ and ψ are formulae, then $\varphi \vee \psi$, $\varphi \wedge \psi$, $\varphi \rightarrow \psi$, $\varphi \equiv \psi$ are all formulae;
(5) only the formulae obtained according to the above Steps (1) to (4) are formulae in $DL(B)$.

The atomic formula (a, v) is also called **attribute-value pair** [7]. If φ is a simple conjunction, which consists of only atomic formulae and connectives $\wedge$, then φ is called a **basic formula**.

For any $x \in U$, the relationship between x and formulae in $DL(B)$ is defined as following:

(1) $x| = (a, v)$ iff $f_a(x) = v$;
(2) $x| = \sim \varphi$ iff not $x| = \varphi$;
(3) $x| = \varphi \wedge \psi$ iff $x| = \varphi$ and $x| = \psi$;
(4) $x| = \varphi \vee \psi$ iff $x| = \varphi$ or $x| = \psi$;
(5) $x| = \varphi \rightarrow \psi$ iff $x| = \sim \varphi \vee \psi$;
(6) $x| = \varphi \equiv \psi$ iff $x| = \varphi \rightarrow \psi$ and $x| = \psi \rightarrow \varphi$.

For formula φ, if $x| = \varphi$, then we say that the object x satisfies formula φ. Let $[\varphi] = \{x \in U | x| = \varphi\}$, which is the set of all those objects that satisfy formula φ. Obviously, formula φ consists of several attribute-value pairs by using connectives. Therefore, $[\varphi]$ is so-called an **attribute-value block** and φ is called the (attribute-value pair) **formula** of the block. For $DL(C)$ and $DL(D)$, they are distinct decision logic languages and have no formulae in common. However, through attribute-value blocks, an association between $DL(C)$ and $DL(D)$ can be established. For example, suppose $\varphi \in DL(C)$ and $\psi \in DL(D)$ and obviously φ and ψ are two different formulae; but if $[\varphi] \subseteq [\psi]$, we can obtain a decision rule $\varphi \to \psi$. Therefore, attribute-value blocks play an important role in acquiring decision rules, especially in acquiring certainty rules.

3 Minimum Rule Sets

Suppose that $\varphi \in DL(C)$ and $\psi \in DL(D)$. Implication form $\varphi \to \psi$ is said to be a (**decision**) **rule** in decision system $(U, C \cup D)$. If both φ and ψ are basic formula, then $\varphi \to \psi$ is called **basic decision rule**. A decision rule is not necessarily useful unless it satisfies some given indices. Below we introduce these indices.

A decision rule usually has two important measuring indices, confidence and support, which are defined as: $conf(\varphi \to \psi) = |[\varphi] \cap [\psi]| / |[\varphi]|, sup(\varphi \to \psi) = |[\varphi] \cap [\psi]| / |U|$, where $conf(\varphi \to \psi)$ and $sup(\varphi \to \psi)$ are **confidence** and **support** of decision rule $\varphi \to \psi$, respectively.

For decision system $DS = (U, C \cup D)$, if rule $\varphi \to \psi$ is true in $DL(C \cup D)$, i.e., for any $x \in Ux | = \varphi \to \psi$, then rule $\varphi \to \psi$ is said to be **consistent** in DS, denoted by $| =_{DS} \varphi \to \psi$; if there exists at least object $x \in U$ such that $x | = \varphi \wedge \psi$, then rule $\varphi \to \psi$ is said to be **satisfiable** in DS. Consistency and satisfiability are the basic properties that must be satisfied by decision rules.

For object $x \in U$ and decision rule r: $\varphi \to \psi$, if $x | = r$, then it is said that rule r **covers** object x, and let **coverage**$(r) = \{x \in U | x | = r\}$, which is the set of all objects that are **covered** by rule r; for two rules, r_1 and r_2, if coverage$(r_1) \subseteq$ coverage(r_2), then it is said that r_2 **functionally covers** r_1, denoted by $r_1 \leq r_2$. Obviously, if there exist such two rules, then rule r_1 is redundant and should be deleted, or in other words, those rules that are functionally covered by other rules should be removed out from rule sets.

In addition, for a rule $\varphi \to \psi$, we say that $\varphi \to \psi$ is **reduced** if $[\varphi] \subseteq [\psi]$ does not hold any more when any attribute-value pair is removed from φ. And this is just known as rule reduction, which will be introduced in next section.

A decision rule set $\wp$ is said to be **minimal** if it satisfies the following properties [3]: (1) any rule in $\wp$ should be consistent; (2) any rule in $\wp$ should be satisfiable; (3) any rule in $\wp$ should be reduced; (4) for any two rules $r_1, r_2 \in \wp$, neither $r_1 \leq r_2$ nor $r_2 \leq r_1$.

In order to obtain a minimum rule set from a given data set, it is required to complete three steps: attribute reduction, rule reduction and rule set minimum. This paper does not introduce attribute reduction methods any more, and we try to propose new methods for rule reduction and for rule set minimum in next sections.

4 Methods of Acquiring Decision Rules

4.1 Rule Reduction

Rule reduction is to keep the minimal attribute-value pairs in a rule such that the rule is still consistent and satisfiable by removing redundant attributes from the rule. For the convenience of discussion, we let $r(x)$ denote a decision rule that is generated with object x, and introduce the following definitions and properties.

Definition 1. For decision system $DS = (U, C \cup D), B = \{a_1, a_2, \ldots, a_m\} \subseteq C$ and $x \in U$, let $\textbf{pairs}(x, B) = (a_1, f_{a_1}(x)) \wedge (a_2, f_{a_2}(x)) \wedge \ldots \wedge (a_m, f_{a_m}(x))$ and let $\textbf{block}(x, B) = [pairs(x, B)] = [(a_1, f_{a_1}(x)) \wedge (a_2, f_{a_2}(x)) \wedge \ldots \wedge (a_m, f_{a_m}(x))]$, and the number m is called the lengths of $pairs(x, B)$ and $block(x, B)$, denoted by $|pairs(x, B)|$ and $|block(x, B)|$, respectively.

Property 1. Suppose $B_1, B_2 \subseteq C$ with $B_1 \subseteq B_2$, then $block(x, B_2) \subseteq block(x, B_1)$.

The proof of Property 1 is straightforward. According to this property, for an attribute subset B, $block(x, B)$ increases with removing attributes from B, but with the prerequisite that $block(x, B)$ does not "exceed" the decision class $[x]_D$, to which x belongs. Therefore, how to judge whether $block(x, B)$ is still contained in $[x]_D$ or not is crucial for rule reduction.

Property 2. For decision system $DS = (U, C \cup D)$ and $B \subseteq C, block(x, B) \subseteq [x]_D$ $(= block(x, D))$ if and only if $f_d(y) = f_d(x)$ for all $y \in block(x, B)$.

The proof of Property 2 is also straightforward. This property shows that the problem of judging whether $block(x, B)$ is contained in $[x]_D$ becomes that of judging whether $f_d(y) = f_d(x)$ for all $y \in block(x, B)$. Evidently, the latter is much easier than the former. Thus, we give the following algorithm for reducing a decision rule.

Algorithm 1: an algorithm for reducing a decision rule
Input: decision system $(U, C \cup D)$ and object $x \in U$, where $C = \{a_1, a_2, \ldots, a_n\}$, $n = |C|, D = \{d\}$
    Output: reduced decision rule $r(x)$
    Begin
        Step 1.   Let $B = C$;
        Step 2.   For $i = 1$ to $|C|$ do
        Step 3.   Let $B = B - \{a_i\}$;
        Step 4.   Compute $block(x, B)$;
        Step 5.   For each $y \in block(x, B)$ do
        Step 6.   If $f_d(y) \neq f_d(x)$ then $\{$ Let $B = B \cup \{a_i\}$; break; $\}$
        Step 7.   Let $\varphi = pairs(x, B)$ and $\psi = (d, f_d(x))$;
        Step 8.   Let $r(x) = \varphi \rightarrow \psi$;
        Step 9.   return $r(x)$;
    End.

The time-consuming step in this algorithm is to compute $block(x, B)$, whose comparison number is $|U||B|$. Therefore, the complexity of this algorithm is $O(|U||C|^2)$ in the worst case. According to Algorithm 1, it is guaranteed at any time that $block(x, B) \subseteq [x]_D = block(x, D)$, so the confidence of rule $r(x)$ is always equal to 1.

4.2 Minimum of Decision Rule Sets

Using Algorithm 1, each object in U can be used to generate a rule. This means that after reducing rules, there are still $|U|$ rules left. Obviously, there must be many rules that are covered by other rules, and hereby we need to delete those rules which are covered by other rules.

For decision system $(U, C \cup D)$, after using Algorithm 1 to reduce each object $x \in U$, all generated rules $r(x)$ constitute a rule set, denoted by RS, i.e., $RS = \{r(x) | x \in U\}$. Obviously, $|RS| = |U|$. Our purpose in this section is to delete those rules which are covered by other rules, or in other words, to minimize RS such that each of the remaining rules is consistent, satisfiable, reduced, and is not covered by other rules.

Suppose $V_d = \{v_1, v_2, \ldots, v_t\}$. We use decision attribute d to partition U into t attribute-value blocks (equivalence classes): $[(d, v_1)], [(d, v_2)], \ldots, [(d, v_t)]$. Let $U_{v_i} = [(d, v_i)]$, and thus $\bigcup_{i \in \{1,2,\ldots,t\}} U_{v_i} = U$ and $U_{v_i} \cap U_{v_j} = \emptyset$, where $i \neq j, i, j \in \{1, 2, \ldots, t\}$. Accordingly, let $RS_{v_i} = \{r(x) | x \in U_{v_i}\}$, where $i \in \{1, 2, \ldots, t\}$. Obviously, $\{RS_{v_i} | i \in \{1,2,\ldots,t\}\}$ is a partition of RS. According to Algorithm 1, for any $r' \in RS_{v_i}$ and $r'' \in RS_{v_j}$, where $i \neq j$, neither $r' \leq r''$ nor $r'' \leq r'$, because coverage$(r') \subseteq U_{v_i}$ while coverage$(r'') \subseteq U_{v_j}$ and then coverage$(r') \cap$ coverage$(r'') = \emptyset$. This means that a rule in RS_{v_i} does not functionally covers any rule in RS_{v_j}. Thus, we can independently minimize each RS_{v_i}, and the union of all the generated rule subsets is the final minimum rule set that we want.

Let independently consider RS_{v_i}, where $i \in \{1, 2, \ldots, t\}$. For $r(x) \in RS_{v_i}$, if there exists $r(y) \in RS_{v_i}$ such that $r(x) \leq r(y)$ ($r(y)$ functionally covers $r(x)$) , where $x \neq y$, then $r(x)$ should be removed from RS_{v_i}, otherwise it should not. Suppose after removing, the set of all remaining rules in RS_{v_i} is denoted by RS'_{v_i}, and thus we can give an algorithm for minimizing RS_{v_i}, which is described as follows.

Algorithm 2: an algorithm for minimizing RS_{v_i}

Input: $U_{v_i} = \{x_1, x_2, \ldots, x_q\}$ and $RS_{v_i} = \{r(x_1), r(x_2), \ldots, r(x_q)\}$

Output: RS'_{v_i}

Begin

Step 1.   Let $\wp = RS_{v_i}$;

Step 2.   For $j = 1$ to q do $// q = |RS_{v_i}|$

Step 3.   Let $\wp = \wp - \{r(x_j)\}$;

Step 4.   Let flag = 1;

Step 5.   For each $r \in \wp$ do

Step 6.   If $x_j \in$ coverage(r) then { flag = 0; break; }

Step 7.   If flag = 1 then $\wp = \wp \cup \{r(x_j)\}$;

Step 8.   Let $RS'_{v_i} = \wp$;

Step 9.   Return RS'_{v_i};

End.

In Algorithm 2, judging if $x_j \in coverage(r)$ takes at most $|C|$ comparison times. But because all rules in RS_{v_i} have been reduced by Algorithm 1, the comparison number should be much smaller than $|C|$. Therefore, the complexity of Algorithm 2 is $O(q^2 \cdot |C|) = O(|U_{v_i}|^2 \cdot |C|)$ in the worst case.

4.3   An Algorithm for Acquiring Minimum Rule Sets

Using the above proposed algorithms and related attribute reduction algorithms, we now can give an entire algorithm for acquiring a minimum rule set from a given data set. The algorithm is described as follows.

Algorithm 3: an algorithm for acquiring a minimum rule set from a data set
Input: decision system $DS = (U, C \cup D)$
Output: a minimum rule set, *minRS*
Begin

Step 1.	Use an attribute reduction algorithm to find a reduct of *DS*, and suppose the reduct is *R*;
Step 2.	Compute *U/R*, and then select one object in each equivalence class in *U/R* to constitute a new decision system $(U', R \cup D)$;
Step 3.	Reduce each object (rule) in $(U', R \cup D)$ using Algorithm 1, and suppose the obtained rule set is denoted by *RS*;
Step 4.	Use decision attribute set *D* to partition U' into several decision classes: $U'_{v_1}, U'_{v_2}, ..., U'_{v_t}$, and then let $RS_{v_i} = \{r(x) \mid x \in U'_{v_i}\}$, where $i \in \{1,2,...,t\}$;
Step 5.	In turn or in parallel minimize $RS_{v_1}, RS_{v_2}, ..., RS_{v_t}$ using Algorithm 2, and suppose corresponding results are $RS'_{v_1}, RS'_{v_2}, ..., RS'_{v_t}$;
Step 6.	Let $minRS = RS'_{v_1} \cup RS'_{v_2} \cup ... \cup RS'_{v_t}$;
Step 7.	Return *minRS*;

End.

In Algorithm 3, there are three steps used to "evaporating" redundant data: Steps 2, 3, 5. These steps also determine the complexity of the entire algorithm. Actually, the newly generated decision system $(U', R \cup D)$ in Step 2 is completely determined by Step 1, which is attribute reduction and has the complexity of about $O(|C|^2|U|^2)$. The complexity of Step 3 is $O(|U'|^2|C|^2)$ in the worst case. Step 5's complexity is $O(|U'_{v_1}|^2 \cdot |C|) + O(|U'_{v_2}|^2 \cdot |C|) + ... + O(|U'_{v_t}|^2 \cdot |C|)$. Because this step can be performed in parallel, so it can be more efficient under parallel environment. Generally, after attribute reduction, the size of a data set would greatly decrease, i.e., $|U'| \ll |U|$. Therefore, computation time of Algorithm 3 is mainly determined by Step 1, so it has the complexity of $O(|C|^2|U|^2)$ in most cases.

5   Experiment Analysis

This section aims to verify the effectiveness of the proposed methods through experiments. There are four UCI data sets (http://archive.ics.uci.edu/ml/datasets.html) used in our experiments, and they are outlined in Table 1. For missing values, they were replaced with the most frequently occurring value on the corresponding attribute.

We executed Algorithm 3 on the four data sets to obtain minimum rule sets. Suppose that the set of finally obtained decision rules on each data set is denoted by $minRS$. The indices that we are interesting in and their meanings are as follows.

- Number of rules: $|minRS|$, i.e., the number of decision rules in $minRS$
- Average value of support: $\frac{1}{|minRS|} \sum_{r \in minRS} sup(r)$, and $minValue = \min_{r \in minRS}\{sup(r)\}$, $maxValue = \max_{r \in minRS}\{sup(r)\}$
- Average value of confidence: $\frac{1}{|minRS|} \sum_{r \in minRS} conf(r)$
- Evaporation ratio: the ratio of removed items (attribute values) to all items (all attribute values)
- Running time: the running time of Algorithm 3, which includes attribute reduction, rule reduction and minimum of decision rule sets, and this index is measured in seconds.

The experimental results on the four data sets are shown in Table 2.

Table 1. Description of the four data sets.

No.	Data sets	Abbreviation	$|U|$	$|C|$	$|V_d|$
1	Dermatology database	Dermatology	366	34	6
2	Tic-Tac-Toe endgame database	Tic-Tac-Toe	958	9	2
3	Mushroom database	Mushroom	8124	22	2
4	Nursery database	Nursery	12960	8	5

Table 2. Experimental results on the four data sets

Data set	Number of rules	Average value of support (minValue, maxValue)	Average value of confidence	Evaporation ratio	Running time (Sec.)
Dermatology	72	0.0146 (0.0027, 0.1257)	1	0.9794	0.14
Tic-Tac-Toe	176	0.0066 (0.0010, 0.0940)	1	0.9072	0.16
Mushroom	17	0.0689 (0.0010, 0.2166)	1	0.9998	2.37
Nursery	305	0.0031 (0.00008, 0.3333)	1	0.9831	31.34

From Table 2, it can be found that the obtained rule sets on the four data sets all have very high evaporation ratio, and each rule in these rule sets has certain support. Specially, there are averagely 0.0689*8124 = 560 objects supporting each rule in the rule set obtained on Mushroom. This shows that these rule sets have relatively strong generalization ability. Furthermore, the running time of Algorithm 3 on each data set is not long and hereby can be accepted by users. In addition, Algorithm 1 can guarantee at any time that $block(x, B) \subseteq [x]_D = block(x, D)$ for all $x \in U$, so the confidence of each rule is always equal to 1, or in other words all the obtained decision rules are deterministic. All these results demonstrate Algorithm 3 is effective and has better application value.

6   Conclusion

Acquiring decision rules from data sets is an important task in rough set theory. This paper conducted our study through the following three aspects so as to provide an effective granulation method to acquire minimum rule sets. Firstly, we introduced decision logic language and attribute-value block technique. Secondly, we used attribute-value block technique to study how to reduce rules and to minimize rule sets, and then proposed effective algorithms for rule reduction and rule set minimum. Thus, together with related attribute reduction algorithm, the proposed granulation method constituted an effective solution to the acquisition of minimum rule sets, which is a kind classifier and can be used for class prediction. Thirdly, we conducted a series of experiments to show that our methods are effective and feasible.

Acknowledgements. This work is supported by the National Natural Science Foundation of China (No. 61363027), the Guangxi Natural Science Foundation (No. 2015GXNSFAA139292).

References

1. Pawlak, Z.: Rough set. Int. J. Comput. Inf. Sci. **11**(5), 341–356 (1982)
2. Guan, Y.Y., Wang, H.K., Wang, Y., Yang, F.: Attribute reduction and optimal decision rules acquisition for continuous valued information systems. Inf. Sci. **179**(17), 2974–2984 (2009)
3. Meng, Z., Jiang, L., Chang, H., Zhang, Y.: A heuristic approach to acquisition of minimum decision rule sets in decision systems. In: Shi, Z., Wu, Z., Leake, D., Sattler, U. (eds.) IIP VII. IFIP AICT, vol. 432, pp. 187–196. Springer, Heidelberg (2014)
4. Hong, T.P., Tseng, L.H., Wang, S.L.: Learning rules from incomplete training examples by rough sets. Expert Syst. Appl. **22**(4), 285–293 (2002)
5. Leung, Y., Wu, W.Z., Zhang, W.X.: Knowledge acquisition in incomplete information systems: a rough set approach. Eur. J. Oper. Res. **168**(1), 164–180 (2006)
6. Li, J.H., Mei, C.L., Lv, Y.J.: Incomplete decision contexts: approximate concept construction, rule acquisition and knowledge reduction. Int. J. Approximate Reasoning **54**(1), 149–165 (2013)
7. Grzymala-Busse, J.W., Clark, P.G., Kuehnhausen, M.: Generalized probabilistic approximations of incomplete data. Int. J. Approximate Reasoning **55**(1), 180–196 (2014)
8. Patrick, G.C., Grzymala-Busse, J.W.: Mining incomplete data with attribute-concept values and "do not care" conditions. In: IEEE International Conference on Big Data. IEEE (2015)

Collective Interpretation and Potential Joint Information Maximization

Ryotaro Kamimura[✉]

IT Education Center and Graduate School of Science and Technology,
Tokai Univerisity, 4-1-1 Kitakaname, Hiratsuka, Kanagawa 259-1292, Japan
ryo@keyaki.cc.u-tokai.ac.jp

Abstract. The present paper aims to propose a new type of information-theoretic method called "potential joint information maximization". The joint information maximization has an effect to reduce the number of jointly fired neurons and then to stabilize the production of final representations. Then, the final connection weights are collectively interpreted by averaging weights produced by different data sets. The method was applied to the data set of rebel participation among youths. The result show that final weights could be collectively interpreted and only one feature could be extracted. In addition, generalization performance could be improved.

Keywords: Collective interpretation · Generalization · Mutual information maximization · Potentiality · Pseudo-potentiality

1 Introduction

Information-theoretic methods have had much influences on neural computing in many aspects of neural learning [1–7]. Though the information-theoretic methods have aimed to describe relations or dependencies between neurons or between layers, due attention has not been paid to those relations. They have even tried to reduce the strength of relations between neurons [8,9]. For example, they have tried to make individual neurons as independent as possible. In addition, they have tried to make the distribution of neurons' firing as uniform as possible. This is simply because difficulty has existed in taking into account neurons' relations or dependencies.

The present paper aims to describe one of the main relations between neurons, namely, relations between input and hidden neurons, because they play critical roles in improving the performance of neural networks, for example, generalization performance. However, it has been few efforts to describe relations between input and hidden neurons from the information-theoretic points of view. To examine relations between input and hidden neurons, we introduce the joint probability between input and hidden neurons. Then, the joint information contained between input and hidden neurons is also introduced. When this joint

© IFIP International Federation for Information Processing 2016
Published by Springer International Publishing AG 2016. All Rights Reserved
Z. Shi et al. (Eds.): IIP 2016, IFIP AICT 486, pp. 12–21, 2016.
DOI: 10.1007/978-3-319-48390-0_2

information increases, only a small number of joint input and hidden neurons fire strongly, while all the others cease to do so.

However, one of the major problems to realize the joint information lies in difficulty in computation. As has been well known, the majority of the information-theoretic methods have this problem of difficulty in computation [7]. To overcome the problem, we have introduced the potential learning [10–13]. In the method, information maximization can be translated into potentiality maximization where a specific neuron is forced to have the largest potentiality to deal with many different situations. Applying the potentiality to joint neurons, potentiality maximization corresponds to a situation where a small number of joint neurons are forced to have larger potentiality.

In addition, the present method aims to propose a new method to interpret final representations. As has been well known, the black-box property of neural networks have prevented them from being applied to practical problems, because in practical applications, the interpretation of final results can be more important than the generalization performance. Usually, neural networks produce completely different types of connection weights, depending on different data sets and initial conditions. The joint information maximization can be used to explain the final representations clearly. When the joint information increases, the number of activated neurons diminishes, which constraints severally the production of many different types of weights. Thus, a few typical connection weights are produced by the joint information maximization. Then, we can interpret those connection weights by averaging them. This type of interpretation is called "collective interpretation" in the present paper. As generalization performance is evaluated in terms of the average values, the interpretation performance can be evaluated collectivity by taking into account all the connection weights produced by diffident data sets and initial conditions.

2   Theory and Computational Methods

2.1   Concept of Joint Information Maximization

Figure 1 shows a concept of joint information maximization. For a data set, when the joint information is maximized, only one joint hidden and input neuron fire strongly with a strong connection weight in Fig. 1(b). For another data set, another joint hidden and input neuron strongly fire in Fig. 1(c). For interpretation, connection weights produced by all data sets are taken into account by averaging connection weights with due consideration for hidden-output connection weights in Fig. 1(e).

2.2   Potential Joint Information Maximization

Potential joint information is based on the potentiality so far defined for hidden neurons [10–13]. As shown in Fig. 1(b), let w_{jk}^t denote connection weights from

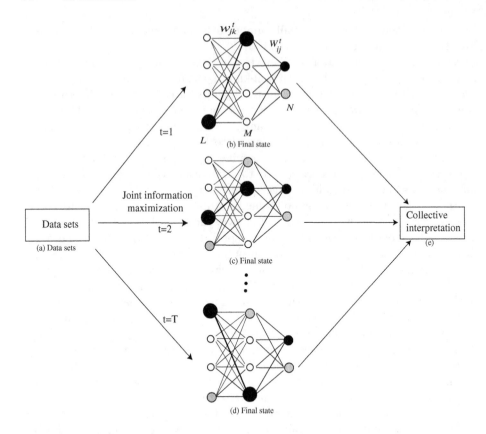

Fig. 1. Concept of joint information maximization with collective interpretation.

the kth input neuron to the jth hidden neuron for the tth data set, then the potentiality v_{jk}^t is defined by

$$v_{jk}^t = \left(w_{jk}^t - w^t\right)^2, \tag{1}$$

where w^t denotes the average weight defined by

$$w^t = \frac{1}{ML} \sum_{j=1}^{M} \sum_{k=1}^{L} w_{jk}^t, \tag{2}$$

where M and L denotes the number of hidden and input neurons. Then, the potentiality is normalized as

$$p(j,k|t) = \frac{v_{jk}^t}{\sum_{m=1}^{M} \sum_{l=1}^{L} v_{ml}^t}. \tag{3}$$

Then, we have the potential joint information

$$PJI = -\sum_{j=1}^{M}\sum_{k=1}^{L} p(j,k) \log p(j,k) + \sum_{t=1}^{T} p(t) \sum_{j=1}^{M}\sum_{k=1}^{L} p(j,k|t) \log p(j,k|t), \quad (4)$$

where T is the number of data sets, $p(t)$ is the probability with which the tth data set is given and

$$p(j,k) = \sum_{t=1}^{T} p(t)p(j,k|t). \quad (5)$$

2.3 Computing Pseudo-Potential Joint Information Maximization

It is possible to differentiate the joint information to have update rules, but much simpler methods have been developed in the name of potential learning. In the method, potentiality maximization is replaced by pseudo-potentiality maximization, which is easily maximized just by changing the parameter. Now, the pseudo-potentiality is defined by

$$\phi_{jk}^{t,r} = \left(\frac{v_{jk}^{t}}{v_{max}^{t}} \right)^{r}, \quad (6)$$

where $r \geq 0$ denotes the potential parameter v_{max} is the maximum potentiality. By normalizing this potentiality, we have the pseudo-firing probability

$$p(j,k|t;r) = \frac{\phi_{jk}^{t,r}}{\sum_{m=1}^{M}\sum_{l=1}^{L}\phi_{ml}^{t,r}}. \quad (7)$$

Then, we have pseudo-information

$$PPJI = -\sum_{j=1}^{M}\sum_{k=1}^{L} p(j,k;r) \log p(j,k;r)$$

$$+ \sum_{t=1}^{T} p(t) \sum_{j=1}^{M}\sum_{k=1}^{L} p(j,k|t;r) \log p(j,k|t;r). \quad (8)$$

The pseudo-information can be increased just by increasing the parameter r, and the joint information can be increased by assimilating pseudo-potentiality $\phi_{jk}^{t,r}$ repeatedly, while the potential parameter increased gradually. The new weights $^{new}w_{jk}^{t}$ are obtained by weighting the old weights $^{old}w_{jk}^{t}$ by the pseudo-potentiality

$$^{new}w_{jk}^{t} = {}^{old}w_{jk}^{t}\, \phi_{jk}^{t,r}. \quad (9)$$

Then, new learning starts with those connection weights as initial ones. This process repeats itself for a fixed number of learning steps.

3   Results and Discussion

3.1   Experimental Outline

The data set was made to infer the probability of rebel participation among youths in the Niger Delta [14]. The number of input patterns was 1,340, and 19 input variables were used. The number of patterns for modeling neural networks was 1000 and the remaining 340 was exclusively for testing. With less than 1000 patterns, improved generalization performance was not obtained by the present and conventional methods. Of 1000 modeling data, 700 training data were randomly and repeatedly taken and ten training sets were prepared. The remaining 300 were used for the early stopping and checking the data sets. The potential parameter r was gradually increased from zero in the first learning step to one in the tenth learning step (final step).

3.2   Mutual Information

Figure 2 shows the joint information as a function of the number of steps. The joint information was simplified by supposing the uniform distribution

$$PJI = \log MN + \frac{1}{T} \sum_{t=1}^{T} \sum_{j=1}^{M} \sum_{k=1}^{L} p(j, k|t) \log p(j, k|t). \tag{10}$$

The information increased gradually and close to 0.6. Though the joint information could be further increased, generalization errors increased in direct proportion to this information increase beyond this point. The results show that the present method can increase the joint information sufficiently.

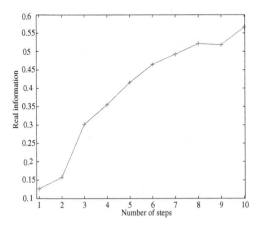

Fig. 2. Potential joint information with 10 hidden neurons for the rebel data set.

3.3   Connection Weights

Figure 3 shows connection weights for the rebel data set when the number of steps increased from one to ten. When the number of steps was one, almost random weights could be seen in Fig. 3(a). When the number of steps was increased from

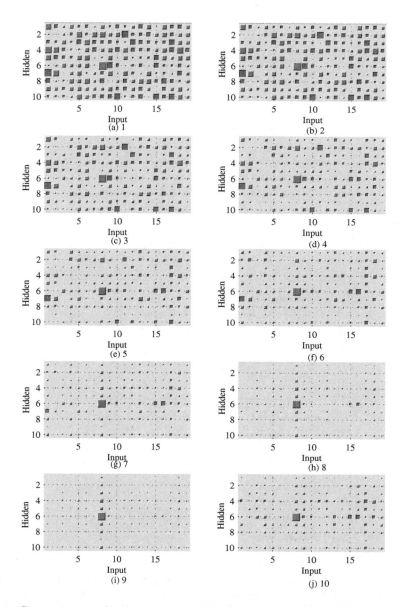

Fig. 3. Connection weights from input to hidden neurons with 10 hidden neurons for the rebel data set. Green and red weights represent positive and negative ones. (Color figure online)

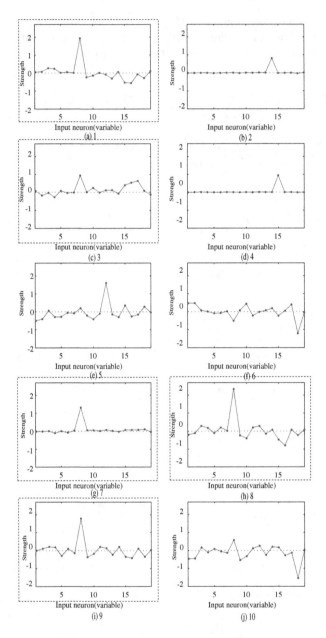

Fig. 4. Adjusted connection weights for ten different data sets from input to hidden neurons with 10 hidden neurons for the rebel data set. Green and red weights denote positive and negative ones. (Color figure online)

two in Fig. 3(b) to six in Fig. 3(f), gradually the number of strong connection weights decreased. Then, when the number of steps was increased from seven in Fig. 3(g) to ten in Fig. 3(j), only one connection weight from the eighth input neuron to sixth hidden neuron became the strongest, while all the other weights became close to zero.

Figure 4 shows adjusted connection weights for the maximum potential hidden neurons j^* by ten different data sets randomly taken from the modeling data set. Adjusted weights for interpretation $c^t_{j^*k}$ was computed by

$$c^t_{j^*k} = \text{sign}(W^t_{1j^*})w^t_{j^*k}, \tag{11}$$

where $\text{sign}(W_{1j^*})$ denote the sign of the weight from the maximum potential hidden neuron to the first output neuron, representing that the youths do not want to participate in the rebel force. As shown in the figure, five out of ten results showed that the input neuron No. 8 had stronger weights than any other ones. Thus, the input neuron No. 8 was collectively considered to be important by the present method.

Figure 5 shows the average connection weights. The average weights were computed by

$$\bar{c}_{j*k} = \frac{1}{T}\sum_{t=1}^{T} c^t_{j^*k} \tag{12}$$

As can be seen in the figure, the input neuron No. 8 had the largest connection weight. The variable No. 8 represents the government's presence in the community in terms of the number of government establishments. Thus, when the government's presence becomes more visible, the youths do not want to participate in the rebel force.

Figure 6 shows the regression coefficients by the logistic regression analysis. In the original data set, a tricky variable was introduced, namely, the variable No. 16 (oil size) and No. 17 (squared oil size), which were naturally correlated, because principally two variables were the same. Thus, they produced the multi-collinearity where two variable responded completely differently to input patterns. On other hand, the present method responded to the two variables almost

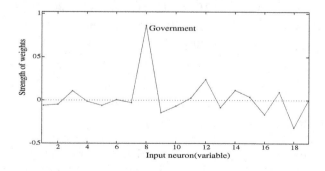

Fig. 5. Collective and average weights for the rebel data set.

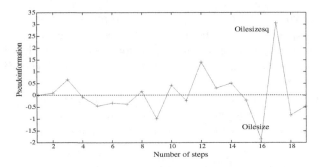

Fig. 6. Regression coefficients for the rebel data set.

evenly. The results show that the present method is good at dealing with this kind of data set with strong correlation between variables. Finally, the interesting thing to note is that except the variables No. 8, No. 16 and No. 17, quite similar weights and coefficients were produced by both methods.

3.4 Generalization Performance

The present method produced the best performance of generalization, comparing with that by the other two conventional methods. Table 1 shows generalization performance by three methods. As can be seen in the table, the best generalization error of 0.1662 on average was obtained by the present method. In addition, the best minimum and maximum error of 0.1382 and 0.2 were obtained by the present method. The second best one was obtained by the BP with the early stopping. Finally, the worst one was obtained by the logistic regression analysis.

Table 1. Summary of experimental results on generalization performance for the rebel data set. The BP(ES) represents the BP with early stopping. The bold face numbers show the best values.

Method	Step	Hidden	Average	Std dev	Min	Max	Inf
Joint	6	10	**0.1662**	0.0181	**0.1382**	**0.2000**	0.4647
BP(ES)	1	10	0.1788	0.0338	**0.1382**	0.2529	0.1262
Logistic			0.2106	0.0129	0.1853	0.2294	

4 Conclusion

The present paper proposed a new information-theoretic method called "joint information maximization". The joint information represents relations between input and hidden neurons. When the joint information increases, the number of strongly connected hidden and input neurons decreases gradually. The method was applied to the rebel participation data set. The results show that the joint information could be increased by the present method. Final results could be interpreted collectively by averaging the connection weights. Finally, generalization performance was improved by the present method. The present method was much simpler than any other conventional information-theoretic methods because of the potential learning. Thus, it can be applied to large-scale and practical problems.

References

1. Linsker, R.: Self-organization in a perceptual network. Computer **21**(3), 105–117 (1988)
2. Barlow, H.B.: Unsupervised learning. Neural Comput. **1**(3), 295–311 (1989)
3. Deco, G., Finnoff, W., Zimmermann, H.: Unsupervised mutual information criterion for elimination of overtraining in supervised multilayer networks. Neural Comput. **7**(1), 86–107 (1995)
4. Bell, A.J., Sejnowski, T.J.: An information-maximization approach to blind separation and blind deconvolution. Neural Comput. **7**(6), 1129–1159 (1995)
5. Linsker, R.: Improved local learning rule for information maximization and related applications. Neural Netw. **18**(3), 261–265 (2005)
6. Principe, J.C., Xu, D., Fisher, J.: Information theoretic learning. Unsupervised Adapt. Filtering **1**, 265–319 (2000)
7. Principe, J.C.: Information Theoretic Learning: Renyi's Entropy and Kernel Perspectives. Information Science and Statistics. Springer, Heidelberg (2010)
8. Comon, P.: Independent component analysis: a new concept. Signal Process. **36**, 287–314 (1994)
9. Bell, A., Sejnowski, T.J.: An information-maximization approach to blind separation and blind deconvolution. Neural Comput. **7**(6), 1129–1159 (1995)
10. Kamimura, R.: Self-organizing selective potentiality learning to detect important input neurons. In: 2015 IEEE International Conference on Systems, Man, and Cybernetics (SMC), pp. 1619–1626. IEEE (2015)
11. Kamimura, R., Kitajima, R.: Selective potentiality maximization for input neuron selection in self-organizing maps. In: 2015 International Joint Conference on Neural Networks (IJCNN), pp. 1–8. IEEE (2015)
12. Kamimura, R.: Supervised potentiality actualization learning for improving generalization performance. In: Proceedings on the International Conference on Artificial Intelligence (ICAI), p. 616, The Steering Committee of the World Congress in Computer Science, Computer Engineering and Applied Computing (WorldComp) (2015)
13. Kitajima, R., Kamimura, R.: Simplifying potential learning by supposing maximum and minimum information for improved generalization and interpretation. In: 2015 International Conference on Modelling, Identification and Control (IASTED 2015) (2015)
14. Oyefusi, A.: Oil and the probability of rebel participation among youths in the niger delta of Nigeria. J. Peace Res. **45**(4), 539–555 (2008)

A Novel Locally Multiple Kernel k-means Based on Similarity

Shuyan Fan[1,2], Shifei Ding[1,2(✉)], Mingjing Du[1,2], and Xiao Xu[1,2]

[1] School of Computer Science and Technology,
China University of Mining and Technology, Xuzhou 221116, China
dingsf@cumt.edu.cn
[2] Key Laboratory of Intelligent Information Processing,
Institute of Computing Technology, Chinese Academy of Sciences,
Beijing 100190, China

Abstract. Most of multiple kernel clustering algorithms aim to find the optimal kernel combination and have to calculate kernel weights iteratively. For the kernel methods, the scale parameter of Gaussian kernel is usually searched in a number of candidate values of the parameter and the best is selected. In this paper, a novel multiple kernel k-means algorithm is proposed based on similarity measure. Our similarity measure meets the requirements of the clustering hypothesis, which can describe the relations between data points more reasonably by taking local and global structures into consideration. We assign to each data point a local scale parameter and combine the parameter with density factor to construct kernel matrix. According to the local distribution, the local scale parameter of Gaussian kernel is generated adaptively. The density factor is inspired by density-based algorithm. However, different from density-based algorithm, we first find neighbor data points using k nearest neighbor method and then find density-connected sets by union-find set method. Experiments show that the proposed algorithm can effectively deal with the clustering problem of datasets with complex structure or multiple scales.

Keywords: Multiple kernel clustering · Kernel k-means · Similarity measure · Clustering analysis

1 Introduction

Unsupervised data analysis using clustering algorithms provides a useful tool. The aim of clustering analysis is to discover the hidden data structure of a dataset according to a certain similarity criterion such that all the data points are assigned into a number of distinctive clusters where points in the same cluster are similar to each other, while points from different clusters are dissimilar [1]. Clustering has been applied in a variety of scientific fields such as web search, social network analysis, image retrieval, medical imaging, gene expression analysis, recommendation systems and market analysis and so on.

Kernel clustering method can handle data sets that are not linearly separable in input space [2], thus, usually perform better than the Euclidean distance based

© IFIP International Federation for Information Processing 2016
Published by Springer International Publishing AG 2016. All Rights Reserved
Z. Shi et al. (Eds.): IIP 2016, IFIP AICT 486, pp. 22–30, 2016.
DOI: 10.1007/978-3-319-48390-0_3

clustering algorithms [3]. Due to simplicity and efficiency, kernel k-means has become a hot research topic. The kernel function is used to map the input data into a high-dimensional feature space, which makes clusters that are not linearly separable in input space become separable. A single kernel is sometimes insufficient to represent the data. Recently, multiple kernel clustering has gained increasing attention in machine learning. Huang et al. propose a multiple kernel fuzzy c-means [4]. By incorporating multiple kernels and automatically adjusting the kernel weights, ineffective kernels and irrelevant features are not crucial for kernel clustering. Zhou et al. use the maximum entropy method to regularize the kernel weights and decide the important kernels [5]. Gao applies multiple kernel fuzzy c-means to optimize clustering and presented mono-nuclear kernel function which is a set of Gaussian kernel function combination assigned different weights resolution [6]. Lu et al. applies multiple kernel k-means clustering algorithm into SAR image change detection [7]. They fuse various features through a weighted summation kernel by automatically and optimally computing the kernel weights, which leads to computational burden. Zhang et al. propose a locally multiple kernel clustering which assigns to each cluster a weight vector for feature selection and combines it with a Gaussian kernel to form a unique kernel for the corresponding cluster [8]. They search the scale parameter of Gaussian kernel by running their clustering algorithm repeatedly for a number of values of the parameter and selecting the best one. Tzortzis et al. overcome the kernel selection problem of maximum margin clustering by employing multiple kernel learning to jointly learn the kernel and a partitioning of the instances [9]. Yu et al. propose an optimized kernel k-means clustering which optimizes the cluster membership and kernel coefficients based on the same Rayleigh quotient objective [10]. Lu et al. improve kernel evaluation measure based on centered kernel alignment and their algorithm needs to be given the initial kernel fusion coefficients [11]. Although the above methods extend from different clustering algorithms, they all employ the alternating optimization technique to solve their extended problems. Specifically, cluster labels and kernel combination coefficients are alternatively optimized until convergence.

Our algorithm is proposed from perspective of similarity measure by calculating a local scale parameter for each data point, which can reflect local distribution of datasets. In addition, another parameter named density factor is introduced in Gaussian kernel function which can describe global structure of data set and avoid kernel k-means running into local optimum. Based on improved similarity measure, our algorithm has several advantages. First, as a kernel method, it has unusual ability in dealing with datasets with multiple scales. Second, it fuses automatically and optimally local and global structures of datasets. Furthermore, our algorithm does not need a good deal of iterations and calculate kernel weights until convergence.

The remainder of this paper is organized as follows: in Sect. 2 we introduce the related works. In Sect. 3 we give a detailed description of our algorithm. Section 4 presents the experimental results and evaluation of our algorithm. Finally, we conclude the paper in Sect. 5.

2 Related Work

2.1 Kernel K-Means

Girolami first proposed the kernel k-means clustering method. It first maps the data points from the input space to higher dimensional feature space through a nonlinear transformation $\phi(\cdot)$ and then minimizes the clustering error in that feature space [12].

Let $D = \{x_1, x_2, \ldots, x_n\}$ be the data set of size n, k be the number of clusters required. The final partition of the entire data set is $\Pi_D = \{C_1, C_2, \ldots, C_k\}$. The objective function is to minimize the criterion function:

$$J = \sum_{j=1}^{k} \sum_{x_i \in C_j} \| \phi(x_i) - m_j \|^2 \tag{1}$$

Where m_j is the mean of cluster C_j. That is

$$m_j = \sum_{x_i \in C_j} \frac{\phi(x_i)}{|C_j|} \tag{2}$$

in the induced space.

$$\left\| \phi(x_i) - m_j \right\|^2 = \left\| \phi(x_i) - \sum_{x_i \in C_j} \frac{\phi(x_i)}{|C_j|} \right\|^2$$
$$= \phi(x_i) \cdot \phi(x_i) + \frac{2}{C_j} \sum_{x_l \in C_j} \phi(x_l) \cdot \phi(x_i) + \frac{1}{|C_j|^2} \sum_{x_l \in C_j} \sum_{x_s \in C_j} \phi(x_l) \cdot \phi(x_s) \tag{3}$$
$$= \kappa(x_i, x_i) + \frac{2}{C_j} \sum_{x_l \in C_j} \kappa(x_i, x_l) + \frac{1}{|C_j|^2} \sum_{x_l \in C_j} \sum_{x_s \in C_j} \kappa(x_l, x_s)$$

Further, $\| \phi(x_i) - m_j \|^2$ can be calculated without knowing the transformation $\phi(\cdot)$ explicitly as formula (3).

Thus, only inner products are used in the computation of the Euclidean distance between a point and a centroid. If given a kernel matrix κ, where $\kappa_{ij} = \phi(x_i) \cdot \phi(x_j)$, A kernel function is commonly used to map the original points to inner products. Given a data set, the kernel k-means clustering has the following steps:

Algorithm1. Kernel k-means clustering algorithm.

Input:
Data set: $D = \{x_1, x_2, \ldots, x_n\}$
The number of cluster: k
Output:
The final Partition: $\Pi_D = \{C_1, C_2, \ldots, C_k\}$.
Method:
Step 1: Initialize the k clusters $\{C_1^0, C_2^0, \ldots, C_k^0\}$
Step 2: For each point x_i and each cluster C_i, compute $\| \phi(x_i) - m_j \|^2$ using formula (3) and assign x_i to its nearest center.
Step 3: Update m_j, for j=1 to k using formula (2).
Step 4: Repeat step 2 and 3 till convergence

2.2 Multiple Kernel k-means

Weighted summation kernel is a common tool for multiple kernel learning. Huang et al. propose multiple kernel k-means algorithm by incorporating weighted summation kernel into the kernel k-means, which results in the multiple kernel k-means algorithm [4]. The MKKM algorithm is solved by updating iteratively the kernel weights. Its objective function is to minimize

$$J_M = \sum_{j=1}^{k} \sum_{x_i \in C_j} \sum_{m}^{M} w_k^2 \parallel \phi_k(x_i) - v_c \parallel^2 \tag{4}$$

$$w_m = \frac{\frac{1}{\beta_m}}{(\frac{1}{\beta_1} + \frac{1}{\beta_2} + \ldots + \frac{1}{\beta_M})}, \beta_m = \sum_{j=1}^{k} \sum_{x_i \in C_j} \parallel \phi(x_i) - m_j \parallel^2$$

Where $\{\phi_k\}_{m=1}^{M}$ are the mapping functions corresponding to multiple kernel functions. $w_m(m = 1, 2, .., M)$ are kernel weights.

3 Locally Multiple Kernel k-means

3.1 Similarity Measure

Selecting a suitable method of similarity measure in cluster analysis is crucial, and it is used as the basis for division [13]. To handle the dataset with multiple scales, we calculate a local scaling parameter σ_i for each data point s_i. The selection of the local scale σ_i can be done by studying the local statistics of the neighborhood of point s_i. s_K is the K'th neighbor of point s_i.

$$\sigma_i = d(s_i, s_K)$$

According to the conception of clustering hypothesis, the data point of intra-class should locate in high-density region, and the data point of inter-class should be separated by low-density region [14]. In order to better describe global structure of data set and avoid kernel k-means running into local optimum, density factor ρ is introduced to discover clusters of arbitrary shape. Combined ρ with formula (6), we propose a new similarity measure as follows:

$$S_{ij} = \exp(\frac{-d^2(s_i, s_j)}{\sigma_i \sigma_j \rho_{ij}}) \tag{5}$$

Density factor is obtained by a simple and powerful way. First, find k neighbor points for each point by k nearest neighbor algorithm and then use k nearest neighbor graph to depict the local neighborhood relation between data points. The neighborhood of a point p is denoted by $N(p)$. For a sample point q, if $q \in N(p)$, we think q is directly density-reachable from point p. Given a sample set D = $\{p_1, p_2, \ldots, p_n\}$, supposed that p_i is directly density-reachable from point p_{i+1}, p_1 is density-reachable from p_n.

If there is a point o such that both, p and q are density-reachable from o, we consider that the point p is density-connected to a point q. Finally, according to all directly density-reachable data points, find density-connected sets by union-find set method which is a very sophisticated and practical data structures and mainly used for processing the merger of the problem of some disjoint sets [15]. Let ρ_{ij} denote density factor between the point s_i and s_j, as follows:

$$\rho_{ij} = \begin{cases} 1, & \text{if } s_i, s_j \text{ are in the same density} - \text{connected set} \\ 0, & \text{otherwise} \end{cases} \tag{6}$$

3.2 Algorithm

From the perspective of similarity measure, we propose a novel locally multiple kernel k-means algorithm (LMKKM). Its basic idea is: firstly, calculate the local scale parameter σ and density factor ρ; subsequently, construct kernel matrix based on our proposed similarity measure; finally, according to the kernel matrix, cluster dataset by kernel k-means. The detail steps of our algorithm are as follows.

Algorithm2. Locally multiple kernel k-means

Input:
  Data set: $D = \{x_1, x_2, \dots, x_n\}$
  The number of cluster: k
Output:
    The final Partition: $\Pi_D = \{C_1, C_2, \dots, C_k\}$.
Method:
Step 1: Calculate distance matrix D using Euclidean distance.
Step 2: Calculate the k nearest neighborhoods and get k nearest neighbor graph.
Step 3: Find density-connected sets by union-find set method.
Step 4: According to density-connected sets, density factor is obtained by formula (6).
Step 5: Calculate similarity matrix according to formula (5).
Step 6: Initialize the k clusters $\{C_1^0, C_2^0, \dots, C_k^0\}$.
Step 7: For each point x_i and each cluster C_i, compute

$$d(x_i, m_j) = \kappa(x_i, x_i) + \frac{2}{C_j} \sum_{x_l \in C_j} \kappa(x_i, x_l) + \frac{1}{|C_j|^2} \sum_{x_l \in C_j} \sum_{x_s \in C_j} \kappa(x_l, x_s)$$

and assign x_i to its nearest center.
Step 8: Update m_j, for j=1 to k using formula (2).
Step 9: Repeat step 7 and 8 till convergence.

Suppose n is the total number of points on the data set. Our local multiple kernel k-means contains three main components: calculating parameters, constructing similarity matrix, and clustering. In the phase of calculating parameters, the complexity of k nearest neighbor algorithm and union-find set method both are $O(n)$. The complexity in calculating the similarity matrix is $O(n^2)$. At the last clustering phase, the complexity of k-means is $O(n)$. Our algorithm does not increase the complexity of kernel k-means, but it improves performance.

4   Experiments

4.1   Artificial Data Clustering

In order to verify the effectiveness of our improved algorithm,we choose three artificial data sets, "smile face", "four lines" and "blobs and circle" to perform an experiment and compare with kernel k-means (KKM) algorithm.

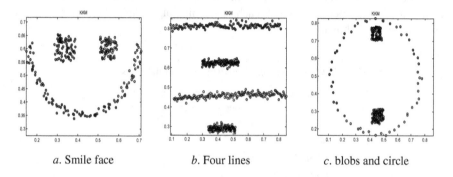

a. Smile face   *b.* Four lines   *c.* blobs and circle

Fig. 1. KKM algorithm's clustering results on artificial data sets

Figure 1 shows KKM algorithm's clustering results on artificial data sets. The scale parameter of Gaussian kernel function is set to be 1 experientially. KKM algorithm measures the similarity between points based on Euclidean distance which can not reflect the intrinsic structure of dataset. Thus, KKM can only gather the similar points in local region into a cluster, but does not satisfy the global coherence hypothesis of the clustering or recognize complex manifold structure of dataset.

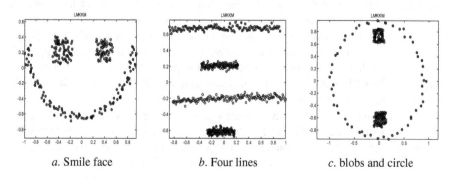

a. Smile face   *b.* Four lines   *c.* blobs and circle

Fig. 2. LMKKM algorithm's clustering results on artificial datasets

Figure 2 shows LMKKM algorithm's clustering results on artificial datasets. It calculates the kernel matrix by formula (6). After the similarity measure involves density factor, it meets both the local coherence hypothesis and global coherence hypothesis of the clustering. Through the approach, the intra-class data points become more compact and the inner-class data points are more discrete.

4.2   Clustering Results

In this subsection, our method is compared with three baseline methods including kernel k-means (KKM), Self-Tuning Spectral Clustering (SSC) [16], locally adaptive multiple kernel clustering (LAMKC) [8]. SSC is a locally adaptive spectral clustering algorithm. LAMKC is a newly proposed multiple kernel clustering algorithm extending form kernel k-means. We carry out experiments on seven UCI datasets. These datasets are often used to test performance of machine learning algorithm. The characteristics of these data sets are shown in Table 1.

Table 1. Data characteristics of real data sets

Data set	Instance number	Attribute number	Cluster number
Iris	150	4	3
Sonar	208	60	2
WDBC	569	30	2
Ionosphere	351	34	2
Yeast	1484	8	10
Dermatology	366	34	6
Zoo	101	16	7

We use the accuracy (*ACC*) to evaluate the clustering performance. Considering the random initialization of clustering centers of kernel k-means, clustering results will be fluctuated. Therefore the clustering experiments are repeated 20 times. Results in first row are the means of the 20 trials, and results in second row (in parentheses) are corresponding standard deviations. The neighborhood size of K is set to 7. For all experiments, the cluster number is set to be the cluster number of each dataset. For LAMKC, the stop condition of the gradient descent method is set to be 0.0001 [8]. In order to compare experimental results fairly, we do not use the Kaufman Approach to select a set of initial centroids in LAMKC. All experiments are conducted on Intel Pentium G2030 CPU with 3.00 GHz processor and 4 G RAM running 64bit-Win7. Clustering results are shown in Table 2.

The results on the seven UCI datasets are shown in Table 2. We use the boldface to mark the best result for each dataset. For the clustering performance measured by *ACC*, the experimental results are encouraging and our algorithm obtained five best results for the seven datasets. Comparable to the kernel k-means and self-tuning spectral clustering, performance of our algorithm is significantly better than that of them on

Table 2. Clustering results on real-world data sets

Data set	KKM	SSC	LAMKC	LMKKM
Iris	0.8980 (0.1204)	0.9067 (0.0004)	0.9127 (0.0407)	**0.9600** (0.0312)
Sonar	0.5505 (0.0054)	0.5192 (0.0048)	0.6319 (0.0038)	**0.7337** (0.0325)
WDBC	0.5211 (0.1012)	0.8682 (0.0001)	**0.8832** (0.1426)	0.8714 (0.2308)
Ionosphere	0.5556 (0.0043)	0.5256 (0.0029)	0.7073 (0.0105)	**0.7620** (0.1704)
Yeast	0.3677 (0.0101)	0.3585 (0.2037)	0.4231 (0.0243)	**0.4502** (0.0041)
Dermatology	0.2268 (0.0164)	0.4426 (0.0004)	**0.6436** (0.1202)	0.5887 (0.0520)
Zoo	0.4109 (0.0589)	0.3953 (0.0184)	0.5208 (0.1331)	**0.6038** (0.0145)

seven datasets. For datasets WDBC and Dermatology, our algorithm is roughly comparable to LAMKC. In most cases, our algorithm can capture structures of datasets and calculate appropriate parameters adaptively while LAMKC searched the parameter of Gaussian kernel in a range. These indicate that our improved similarity measure has the capability to capture local and global structures of datasets with complexity so as that our algorithm can complete the tasks of clustering efficiently.

5 Conclusions

Conventional multiple kernel clustering algorithms aim to construct a global combination of multiple kernels in input space and have to kernel combination coefficients iteratively. In this paper, we proposed a local multiple kernel clustering method based on similarity measure. Our method is dedicated to the datasets with varying local distributions. Instead of using a uniform combination of multiple kernels over the whole input space, our method associates to each data point a localized kernel and combined with density factor simultaneously. Taking local and global structures into consideration, our similarity measure can depict distributed situation of dataset. Results of clustering experiments on artificial datasets and UCI datasets demonstrate that our locally multiple kernel clustering method can deal with datasets with multiple scales and not fall into local optimal.

There are three points remaining for further research. First, the time complexity of our algorithm is the same as kernel k-means's, so it will spend much time when processing big data set. Further study is necessary on how to reduce the time complexity of the algorithm and improve the efficiency of clustering. Second, kernel k-means is sensitive to the initial cluster centers. We can improve kernel k-means from this perspective. Third, following the idea of this paper, we can construct better multiple kernel k-means methods based on the other kernel evaluation measures.

Acknowledgements. This work is supported by the National Natural Science Foundation of China (Nos. 61379101, 61672522), and the National Key Basic Research Program of China (No. 2013CB329502).

References

1. Ding, S., Zhang, J., Jia, H., et al.: An adaptive density data stream clustering algorithm. Cogn. Comput.n **8**(1), 30–38 (2016)
2. Chitta, R.: Kernel-based clustering of big data. Dissertations & Theses – Gradworks (2015)
3. Chitta, R., Jin, R., Havens, T.C., Jain, A.K.: Scalable Kernel Clustering: Approximate Kernel k-means. Eprint Arxiv (2014)
4. Huang, H.C., Chuang, Y.Y., Chen, C.S.: Multiple kernel fuzzy clustering. IEEE Trans. Fuzzy Syst. **20**(1), 120–134 (2012)
5. Zhou, J., Chen, C.L., Chen, L., Maximum-entropy-based multiple kernel fuzzy c-means clustering algorithm. In: IEEE International Conference on Systems, Man and Cybernetics IEEE (2014)
6. Gao, S.: The application of clustering optimization in data mining based on multiple kernel function FCM. J. Comput. Inf. Syst. **11**(11), 3977–3986 (2015)
7. Jia, L., Li, M., Zhang, P., et al.: SAR image change detection based on multiple kernel k-means clustering with local-neighborhood information. IEEE Geosci. Remote Sens. Lett. **13**(6), 1–5 (2016)
8. Zhang, L., Hu, X.: Locally adaptive multiple kernel clustering. Neurocomputing **137**(11), 192–197 (2014)
9. Tzortzis, G., Likas, A.: Ratio-based multiple kernel clustering. In: Calders, T., Esposito, F., Hüllermeier, E., Meo, R. (eds.) ECML PKDD 2014, Part III. LNCS, vol. 8726, pp. 241–257. Springer, Heidelberg (2014)
10. Yu, S., Tranchevent, L., Moor, B.D., et al.: Optimized data fusion for kernel k-means clustering. IEEE Trans. Pattern Anal. Mach. Intell. **35**(5), 1031–1039 (2011)
11. Lu, Y., Wang, L., Lu, J., et al.: Multiple kernel clustering based on centered kernel alignment **47**(11), 3656–3664 (2014)
12. Girolami, M.: Mercer kernel-based clustering in feature space. IEEE Trans. Neural Netw. **13**(3), 780–784 (2002)
13. Yan, J., Cheng, D., Zong, M., Deng, Z.: Improved spectral clustering algorithm based on similarity measure. In: Luo, X., Yu, J.X., Li, Z. (eds.) ADMA 2014. LNCS, vol. 8933, pp. 641–654. Springer, Heidelberg (2014)
14. Jia, H., Ding, S., Meng, L., et al.: A density-adaptive affinity propagation clustering algorithm based on spectral dimension reduction. Neural Comput. Appl. **25**(7–8), 1557–1567 (2014)
15. Kaplan, H., Shafrir, N., Tarjan, R.E.: Union-find with deletions. In: Proceedings of the thirteenth annual ACM-SIAM symposium on Discrete algorithms, pp. 19–28 (2002)
16. Manor, M.L.: Self-tuning spectral clustering. Adv. Neural Inf. Process. Syst. **17**, 1601–1608 (2004)

Direction-of-Arrival Estimation for CS-MIMO Radar Using Subspace Sparse Bayesian Learning

Yang Bin[1](✉), Huang Dongmei[2], and Li Ding[2]

[1] Shanxi Electric Power Technical College, Taiyuan 030021, China
wfqsyyy@163.com
[2] Naval Command College, Nanjing 210016, China

Abstract. We address the problem of direction-of-arrival (DOA) estimation for compressive sensing based multiple-input multiple-output (CS-MIMO) radar. The spatial sparsity of the targets enables CS to be desirable for DOA estimation. By discretizing the possible target angles, a overcomplete dictionary is constructed for DOA estimation. A structural sparsity Bayesian learning framework is presented for support recovery. To improve the recovery accuracy and speed up the Bayesian iteration, a subspace sparse Bayesian learning algorithm is developed. The proposed scheme, which needs less iteration steps, can provides high precision DOA estimation performance for CS-MIMO radar, even at the condition of low signal-to-noise ratio and coherent sources. Simulation results verify the usefulness of our scheme.

Keywords: Multiple-input multiple-output radar · Sparse Bayesian learning · Angle estimation · Subspace decomposition

1 Introduction

Multiple-input multiple-output (MIMO) radar is a relatively new concept for radar system. By exploiting multiple antennas in both transmit and receive end, the extra visual antenna aperture is formed. The visual aperture makes the performance of MIMO radar better than the traditional phased array radar [1]. Generally speaking, the array geometry in MIMO radar can be divided into two categories, the uniform array and the nonuniform array. Elements in the uniform array geometry must be spaced at intervals no larger than half wavelength of the carrier signal thus to avoid phase ambiguity. Typical uniform arrays including linear arrays, uniform circular array and L-shape or rectangle array. The nonuniform array setup is much more flexible than the uniform array configuration [2]. The minimum redundancy linear array and the random array are belong to this kind of array.

Direction-of-arrival (DOA) estimation is a fundamental problem in MIMO radar that has aroused extensive attention. Existing estimation algorithms including Capon [3], multiple signal classification (MUSIC) [4], the estimation method of signal parameters via rotational invariance techniques (ESPRIT) [5–7], the parallel factor analysis (PARAFAC) [8, 9]. However, the majority of the above algorithms are

Z. Shi et al. (Eds.): IIP 2016, IFIP AICT 486, pp. 31–38, 2016.
DOI: 10.1007/978-3-319-48390-0_4

effectiveless with the nonuniform arrays. Algorithm Capon and MUSIC are effective with nonuniform configuration, they only perform well with large number of snapshot. Besides, additional prior information is needed in this algorithm, such as the number of targets, the noise level, et al. Recently, compressive sensing (CS) theory has attracted extensive attention in the field of array signal processing [10–12]. In this paper, we focus on the compressive sensing based MIMO (CS-MIMO) radar [13]. In CS-MIMO radar, transmit and receive elements are randomly placed over a large aperture and spatial sampling is applied at sub-Nyquist rate. The random array setup would achieve similar resolution with significantly fewer elements. In fact, the targets in CS-MIMO can be viewed as sparse in the background. Therefore, DOA estimation can be regarded as a sparse inverse problem from multiple measurement vectors (MMV) in CS. Actually, there are many excellent algorithms for the MMV problem, such as Basis Pursuit (BP) [14], Orthogonal Matching Pursuit (OMP) [15], FOCal Underdetermined System Solver (FOCUSS) [16], Sparse Bayesian Learning (SBL) [17]. The sparse inverse problem makes the DOA estimation more accurate than the traditional methods. Unfortunately, the computational complexity of BP algorithm is too large to engineering implementation. Both OMP and FOCUSS are sensitivity to noise, and it is hard for FOCUSS to choose a proper regularization parameter. SBL is a blind recovery algorithm that always achieves the sparsest global minima. The statistical model that SBL based provides a flexible framework to exploit special structures in the model [18], which may significantly improve the recovery performance. A SBL based DOA estimation algorithm has been proposed for CS-MIMO radar [19], but it suffers from high computational complexity, making it unsuitable for the radar system.

In this paper, a structural correlated subspace SBL algorithm is derived for DOA estimation in CS-MIMO radar. The problem of DOA estimation is formulated as a sparse inverse problem. A statistical model is present which utilize the intra- and the extra- information of the data. DOA estimation is linked to parameters iteration in the SBL model. To speed up the learning process, the subspace-based sparse Bayesian learning (S-SBL) algorithm is developed. The proposed algorithm can reduce the complexity of the recovery process and enable the radar system keep good performance in lower signal-to-noise ratio (SNR), even with coherent sources. Experimental results show the proposed scheme performs better than the existing algorithms.

The paper outline is as follows. The data model for the CS-MIMO radar is presented in Sect. 2. The proposed S-SBL algorithm is derived in Sect. 3. Simulation results are given in Sect. 4. We end the paper by a brief concluding in Sect. 5.

Notation, capital letters and lower case in bold denote, respectively, matrices and vectors. The superscript $(X)^T, (X)^H, (X)^{-1}$ and $(X)^\dagger$ represent the operations of transpose, Hermitian transpose, inverse and pseudo-inverse, respectively; $X_{k\bullet}$ and $X_{\bullet k}$ represent the k-th row and k-th column of X. The subscript $\|X\|_F$ denote the Frobenius norm of X; $\otimes$ stands for the Kronecker product; The Khatri-Rao product (column-wise Kronecker product) is denoted by $\odot$, i.e., $[a_1, \cdots, a_K] \odot [b_1, \cdots, b_K] = [a_1 \otimes b_1, \cdots, a_K \otimes b_K]$; The $M \times M$ identity matrix is denoted by I_M, and the $M \times M$ inverse permutation matrix is denoted by II_M.

2 Signal Model

Consider the model for monostatic CS-MIMO radar in Fig. 1 [13]. The radar system is consist of M transmit elements and N receive elements. Both transmit elements and receive elements are located in the x-axis. The total transmit aperture and the receive aperture is Z_{TX} time and Z_{RX} time of the carrier signal wavelength, respectively. The position of the m-th transmitter and the n-th receiver is denoted by $\frac{Z}{2}\zeta_m$ and $\frac{Z}{2}\xi_n$, respectively. Define $Z = Z_{TX} + Z_{RX}$, with ζ_m lies in the interval $-\left[\frac{Z_{TX}}{2}, \frac{Z_{TX}}{2}\right]$ and ξ_n lies in the interval $\left[-\frac{Z_{RX}}{2}, \frac{Z_{RX}}{2}\right]$. Suppose that the transmit elements' locations ζ are independent and identically distributed (i.i.d.) random variables governed by a distribution $p(\zeta)$. Similarly, the positions of the receivers ξ be drawn i.i.d. from a distribution $p(\xi)$. We assumed that there are K targets appearing in the far-field of the CS-MIMO radar system, the DOA of the k-th target is denoted by θ_k. Supposed that the transmit elements emit ideal orthogonal waveforms. In the receive end, the matched filters are used to separate the information from each visual transmitter-to-receiver path. The output of the matched filters with sampled L snapshots can be expressed as

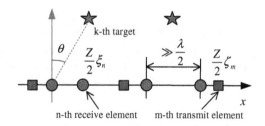

Fig. 1. Signal model of DOA estimation for CS-MIMO radar

$$Y = [A_T(\theta) \odot A_R(\theta)]S + N = AS + N \tag{1}$$

where $A_T(\theta) = [a_t(\theta_1), a_t(\theta_2), \cdots, a_t(\theta_K)] \in \mathbb{C}^{M \times K}$ is the transmit direction matrix, with the k-th $(k = 1, 2, \cdots, K)$ transmit steering vector $a_t(\theta_k)$ is denoted by $a_t(\theta_k) = [\exp(j\pi Z\zeta_1 \sin\theta_k), \exp(j\pi Z\zeta_2 \sin\theta_k) \cdots, \exp(j\pi Z\zeta_M \sin\theta_k)]^T \in \mathbb{C}^{M \times 1}$. $A_R(\theta) = [a_r(\theta_1), a_r(\theta_2), \cdots, a_r(\theta_K)] \in \mathbb{C}^{N \times K}$ is the receive direction matrix, and the k-th $(k = 1, 2, \cdots, K)$ receive steering vector $a_r(\theta_k)$ is given by $a_r(\theta_k) = [\exp(j\pi Z\xi_1 \sin\theta_k), \exp(j\pi Z\xi_2 \sin\theta_k) \cdots, \exp(j\pi Z\xi_N \sin\theta_k)]^T \in \mathbb{C}^{N \times 1}$. The visual direction matrix is $A = [A_T(\theta) \odot A_R(\theta)] \in \mathbb{C}^{MN \times K}$. $S \in \mathbb{C}^{K \times L}$ is the source matrix and $N \in \mathbb{C}^{MN \times L}$ represent an additive Gaussian white noise matrix with variance σ_N^2.

Additional assumption is that the sources are located in the maximum unambiguous angles. By discretizing the possible angle on a fine uniform grid, i.e., $[\varphi_1, \varphi_2 \cdots, \varphi_G]$ $(K \ll G)$, we can obtain an visual overcomplete dictionary $B = [b(\varphi_1), b(\varphi_2) \cdots, b(\varphi_G)] \in \mathbb{C}^{MN \times G}$ for DOA estimation, where the g-th $(g = 1, 2 \cdots, G)$ steering vector

$b(\varphi_g) = a_t(\varphi_g) \otimes a_r(\varphi_g)$, with $a_t(\varphi_g)$ is $a_t(\varphi_g) = [\exp(j\pi Z\zeta_1 \sin \varphi_g), \exp(j\pi Z\zeta_2 \sin \varphi_g) \cdots, \exp(j\pi Z\zeta_M \sin \varphi_g)]^T \in \mathbb{C}^{M \times 1}$, and $a_r(\varphi_g) = [\exp(j2\pi Z\xi_1 \sin \varphi_g), \exp(j2\pi Z\xi_2 \sin \varphi_g) \cdots, \exp(j2\pi Z\xi_M \sin \varphi_g)]^T \in \mathbb{C}^{N \times 1}$. Therefore, the received data in (1) can be written as

$$Y = BX + N, \quad X_{g\bullet} = \begin{cases} S_{k\bullet}, & \text{when } \varphi_g = \theta_k \\ 0, & \text{others} \end{cases}, \quad (2)$$

Noting that $K \ll MN < G$, the model in (2) is a MMV problem [16]. The DOA estimation problem can be expressed as to recover θ from the given measurement matrix Y and the known matrix B, which is equal to recovery the support $supp$ (X) (non-zero rows of X). The previous work in [13] has shown that once MN satisfies certain boundary condition, B would provide a uniform recovery guarantee for (2). Relying on a random array geometry, the model in (2) is linked to the CS framework.

3   Environment Awareness

The subspace that obtained from the covariance matrix decomposition can improve the DOA estimation accuracy in low SNR scene. The covariance matrix of the received data can be estimated by $\hat{R}_Y = YY^H/L$. If the sources are noncoherent, the rank of R_Y is $\text{rank}(R_Y) = K$. But there is a rank loss if the sources are coherent. To maintain the rank of R_Y, the spatial smooth method is applied to R_Y. The estimated covariance matrix becomes

$$R = R_Y + \Pi_{MN} R_Y^* \Pi_{MN} \quad (3)$$

The eigenvalue decomposition of R is

$$R = U\Sigma U^H = \sum_{j=1}^{MN} \lambda_j u_j u_j^H = U_s \Sigma_s U_s^H + U_n \Sigma_n U_n^H \quad (4)$$

where $\Sigma = diag(\lambda_1, \lambda_2, \cdots, \lambda_{MN})$, the eigenvalues are complied with $\lambda_1 \geq \cdots \geq \lambda_K > \lambda_{K+1} = \cdots = \lambda_{MN} = \sigma_N^2$. The eigenvectors $u_1, u_2, \cdots, u_k$ corresponding to the K larger eigenvalues $\lambda_1, \lambda_2, \cdots, \lambda_N$ construct signal subspace $U_s = [u_1, u_2, \cdots, u_k]$, with $\Sigma_s = [\lambda_1, \lambda_2, \cdots, \lambda_K]$. Similarly, the later $MN-K$ eigenvalue are depending on the noise and their numeric values are σ_N^2. The eigenvectors $u_{K+1}, u_{K+2}, \cdots, u_{MN}$ corresponding to $\lambda_{K+1}, \lambda_{K+2}, \cdots, \lambda_{MN}$ construct noise subspace $U_n = [u_{K+1}, u_{K+2}, \cdots, u_N]$, and $\Sigma_n = [\lambda_{K+1}, \lambda_{K+2}, \cdots, \lambda_N]$. Generally, the number of the sources can be estimated by the distribution of the eigenvalue. Let $Z = U_s \in \mathbb{C}^{MN \times K}$ denotes the estimated signal subspace, hence there exist a nonsingular matrix $T \in \mathbb{C}^{MN \times MN}$ that

$$Z = BT \tag{5}$$

Note that the dimension of Z is much smaller than Y, therefore the required iteration steps of SBL for Z will less than Y Let λ be the variance parameter of the noise, γ_g is a hyperparameter, R is the covariance matrix of $X_{g\bullet}$, which captures the intra-block correlation of the source. The mean and covariance parameters are given by μ and Σ. According to [18], the $d + 1$-th iteration process of the structural correlated SBL algorithm can be summarized as follows

$$
\begin{cases}
\left(\gamma_g\right)_{d+1} = \frac{1}{L} \left(X_{g\bullet}\right)_d R_d^{-1} \left(X_{g\bullet}^T\right)_d + \left((\Sigma_x)_d\right)_{gg}, \quad g = 1, 2 \cdots, G \\
R_{d+1} = \tilde{R}_{d+1} / \|\tilde{R}_{d+1}\|_F, \quad \tilde{R}_{d+1} = \sum_{g=1}^G \left(X_{g\bullet}^T\right)_d \left(X_{g\bullet}\right)_d + \eta I \\
\lambda_{d+1} = \frac{1}{MNL} \|Z - BX_d\|_F^2 + \frac{\lambda_d}{G} Tr[BH_d] \\
(\Sigma_x)_{d+1} = \left(\Lambda_d^{-1} + \frac{1}{\lambda_d} B^T \Lambda_d\right), \quad \Lambda = diag[\gamma_1, \gamma_2, \cdots, \gamma_G] \\
X_{d+1} = H_d Z, \quad H_d = \Lambda_d B^T \left(\lambda_d I + B\Lambda_d B^T\right)^{-1}
\end{cases} \tag{6}
$$

where η is a positive scalar. With the iteration of the SBL, the matrixes are trend to stable. The iteration will repeat until algorithm convergence thus we obtain the estimated $\hat{X}$. With the non-zeros rows in $\hat{X}$ we can get the estimated DOAs.

4 Simulation Results

In this section, 1000 Monte Carlo trails are used to assess the DOA estimation performance of the S-SBL algorithm. In our simulation, the successful-rate and average running time are used for performance assessment. A successful trial was recognized if the indexes of estimated sources with K largest l_2-norms were the support of X. Parameters in our simulation are set to $Z = 250$, $M = N = 6$, $\theta = [15, 40, 65]$. The range of the angle in the dictionary is $[0°, 90°]$ with interval $0.1°$. The SNR in the simulation is defined as $10 \log \left(\|Y - N\|_2^2 / \|N\|_2^2\right)$. We compared our algorithm with traditional MUSIC algorithm [4], the OMP method [15], the FOCUSS algorithm [16] and the SBL method [17].

Figures 2 and 3 depict the noncoherent sources estimation performance comparison of all the algorithms with different SNR, respectively. It can be seen from Fig. 2 that all algorithms would achieve better performance with the increasing SNR, while the S-SBL algorithm provides almost precise results. Another result is that the S-SBL runs faster than the SBL method.

Figures 4 and 5 depict the coherent sources estimation performance comparison of all the algorithms with different SNR, respectively. In the simulation, the second target and the third target are coherent. It can be seen from the results that the Bayesian

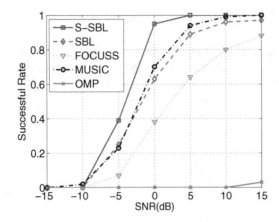

Fig. 2. Successful Rate comparison for noncoherent sources with different SNR

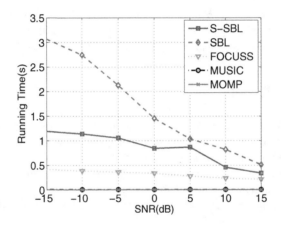

Fig. 3. Average running time comparison for noncoherent sources with different SNR

algorithms are outperform the other algorithms in this situation. Besides, the S-SBL method perform better than the SBL approach, and the S-SBL method requires less computational load than the SBL method.

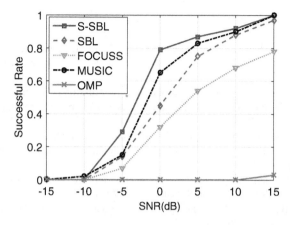

Fig. 4. Successful rate comparison for coherent sources with different SNR

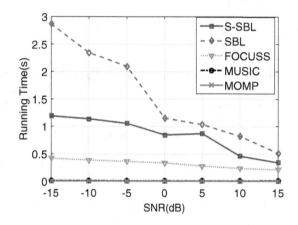

Fig. 5. Average running time comparison for coherent sources with different SNR

5 Conclusions

In this paper, we have proposed a subspace-based SBL algorithm for DOA estimation in CS-MIMO radar. The subspace operation deduces the computational load and improves the estimation accuracy in low SNR scene. Our work links the estimation of angle to the parameters learning in the Bayesian framework. The S-SBL method does not require the prior information of the number of targets, and outperforms the traditional MUSIC algorithm and existing CS recovery methods.

References

1. Fishler, E., Haimovich, A., Blum, R., Chizhik, D., Cimini, L., Valenzuela, R.: MIMO radar: an idea whose time has come. In: Proceeding IEEE Radar Conference, pp. 71–78, April 2004
2. Wen F., Zhang G.: Two-dimensional direction-of-arrival estimation for trilinear decomposition-based monostatic cross MIMO radar. Mathematical Problems in Engineering (2013). Article No. 427980
3. Zhang, X., Huang, Y., Chen, C., Li, J., Xu, D.: Reduced-complexity Capon for direction of arrival estimation in a monostatic multiple-input multiple-output radar. IET Radar Sonar Navig. 6(8), 796–801 (2012)
4. Yan, H., Li, J., Liao, G.: Multitarget identification and localization using bistatic MIMO radar systems. EURASIP J. Adv. Signal Process. 2008, 48 (2008)
5. Duofang, C., Baixiao, C., Guodong, Q.: Angle estimation using ESPRIT in MIMO radar. Electron. Lett. 44(12), 770–771 (2008)
6. Jinli, C., Hong, G., Weimin, S.: Angle estimation using ESPRIT without pairing in MIMO radar. Electron. Lett. 44(24), 1422–1423 (2008)
7. Zheng, Z.D., Zhang, J.Y.: Fast method for multi-target localisation in bistatic MIMO radar. Electron. Lett. 47(2), 138–139 (2011)
8. Zhang, X., Xu, Z., Xu, L., Xu, D.: Trilinear decomposition-based transmit angle and receive angle estimation for multiple-input multiple-output radar. IET Radar Sonar Navig. 5(6), 626–631 (2011)
9. Li, J., Zhou, M.: Improved trilinear decomposition-based method for angle estimation in multiple-input multiple-output radar. IET Radar Sonar Navig. 7(9), 1019–1026 (2013)
10. Wen, F.Q., Tao, Y., Zhang, G.: Analogue-to-information conversion using multi-comparator-based integrate-and-fire sampler. Electron. Lett. 51(3), 246–247 (2015)
11. Wen, F.Q., Zhang, G., Ben, D.: Estimation of multipath parameters in wireless communications using multi-way compressive sensing. J. Syst. Eng. Electron. 26(5), 908–915 (2015)
12. Wen, F.Q., Zhang, G., Ben, D.: Adaptive selective compressive sensing based signal acquisition oriented toward strong signal noise scene. KSII Trans. Internet Inf. Syst. 9(9), 3559–3571 (2015)
13. Rossi, M., Haimovich, A.M., Eldar, Y.C.: Spatial compressive sensing for MIMO radar. IEEE Trans. Sign. Proces. 62(2), 419–430 (2014)
14. Chen, S.S., Donoho, D.L., Saunders, M.A.: Atomic decomposition by basis pursuit. SIAM Rev. 43(1), 129–159 (2001)
15. Tropp, J.A., Gilbert, A.C.: Signal recovery from random measurements via orthogonal matching pursuit. IEEE Trans. Inf. Theory 53(12), 4655–4666 (2007)
16. Cotter, S.F., Rao, B.D., Engan, K., Kreutz-Delgado, K.: Sparse solutions to linear inverse problems with multiple measurement vectors. IEEE Trans. Sign. Proces. 53(7), 2477–2488 (2005)
17. Tipping, M.E.: Sparse Bayesian learning and the relevance vector machine. The J. Mach. Learn. Res. 1, 211–244 (2001)
18. Wen, F.Q., Zhang, G., Ben, D.: A recovery algorithm for multitask compressive sensing based on block sparse Bayesian learning. Acta Phys. Sin. 64(7), 70201 (2015)
19. Wen, F.Q., Zhang, G., Ben, D.: Direction-of-arrival estimation for multiple-input multiple-output radar using structural sparsity Bayesian learning. Chin. Phys. B 24(11), 110201 (2015)

Data Mining

Application of Manifold Learning to Machinery Fault Diagnosis

Jiangping Wang$^{(\boxtimes)}$, Tengfei Duan$^{(\boxtimes)}$, and Lujuan Lei

School of Mechanical Engineering, Xi'an Shiyou University,
Xi'an 710065, Shanxi, China
jpwang@xsyu.edu.cn, tfduan@163.com

Abstract. The essence of machinery fault diagnosis is pattern recognition. Extracting the fault pattern contained in the vibration signal is the frequently used method to diagnose mechanical fault. Manifold Learning is widely used to extract the non-linear structure within the data and could do the dimensionality reduction of high-dimensional signal. Therefore manifold learning is employed to diagnose the machinery fault. The feature space is constructed by characters in time-frequency domain of vibration signal firstly, and then the manifold learning method named as sparse manifold clustering and embedding is used to extract the essential nonlinear structure of feature space. Afterwards, the fault diagnosis is implemented with spectral clustering and support vector machine. The experiment demonstrates that the approach can effectively diagnose the fault of Machinery.

Keywords: Manifold learning · Fault diagnosis · Bearing

1 Introduction

With the increment of precision and operation complexity in industry, the techniques of machinery fault diagnosis, which is used for the safe guarantee in operation of mechanical equipments, has gained more and more attention. At present, data-driven methods play a important role in the mechanical fault diagnosis. The vibration signal contains a wealth of information which indicate the condition of machinery, therefore it is widely used in fault diagnosis that the method based on vibration signal processing [1]. The characters of vibration signal in time-frequency domain that is used to establish the initial feature space can be extracted by methods such as Fourier transform or wavelet transform etc. And then, the classifier is constructed with the further processing of initial feature space [1]. The dimensionality reduction method is taken to get a low-dimensional feature representation since the initial feature space is generally high-dimensional. The frequently used dimensional reduction methods, such as principal component analysis [2] or linear discriminant analysis [3], is the linear dimensional reduction method. For the nonlinear signal, the linear method is not the best choice.

© IFIP International Federation for Information Processing 2016
Published by Springer International Publishing AG 2016. All Rights Reserved
Z. Shi et al. (Eds.): IIP 2016, IFIP AICT 486, pp. 41–49, 2016.
DOI: 10.1007/978-3-319-48390-0_5

Manifold learning has been applied in many fields since three articles published in Science [4–6]. Manifold learning is a nonlinear data dimensional reduction method, which is used to extract low-dimensional manifold structure embed in high-dimensional space. Therefore there is a new way to diagnose the machinery fault. Generally speaking, the operation data of machinery with the same condition lie on the same manifold and the different condition lie on different manifold [7]. Based on this setting, manifold learning can be applied to machinery fault diagnosis.

A typical manifold learning method includes Local Linear Embedding (LLE) [5], Isometric Mapping (ISOMAP) [6] etc. A proper choice of neighborhood size is critical in manifold Learning. Using a fixed neighborhood size is inappropriate, such as in LLE and ISOMAP, because the curvature of the manifold and the density of the sample points may be different in different regions of the manifold in the engineering applications. Sparse Manifold Clustering and Embedding (SMCE) is proposed in [8]. In SMCE, the automatic choice of the neighborhood size is done by solving a sparse optimization problem. Thus SMCE is more suitable for engineering applications. In this paper, we briefly introduce the principle of SMCE and discuss its application in feature extraction and diagnosis of mechanical fault.

2   The principle of SMCE

SMCE can be used for manifold clustering and dimensionality reduction for multiple nonlinear manifold simultaneously. The key difference between SMCE and other manifold learning methods is that SMCE can automatically search neighbors by solving a sparse optimization problem and weight matrix is established by calculating weights between each points and its neighbors. And then, spectral clustering [9] and Laplacian eigenmap [10] can be used for manifold clustering and dimensional reduction respectively.

Given a set of N samples $\{x_i \in \mathbb{R}^D\}_{i=1}^N$ lying in n different manifolds $\{\mathcal{M}_l\}_{l=1}^n$ with intrinsic dimensions $\{d_l\}_{l=1}^n$. For each sample point $x_i \in \mathcal{M}_l$ consider the smallest ball $\mathcal{B}_i \subset \mathbb{R}^D$ contains that the $d_l + 1$ nearest neighbors of x_i. Namely, the affine subspaces with intrinsic dimensions d_l is spanned by the $d_l + 1$ neighbors. This is the fundamental assumption of SMCE. That is,

$$|| \sum_{j \in \mathcal{B}_i} k_{ij}(x_j - x_i)||_2 \leqslant \epsilon \quad and \quad \sum_{j \in \mathcal{B}_i} k_{ij} = 1 \tag{1}$$

Where $x_j \in \mathcal{B}_i$ and $j \neq i$, $\epsilon(\geqslant 0)$ is the upper bound of error, k_{ij} is coefficient. It is hard to know the diameter of $\mathcal{B}_i$, therefore the Eq. (1) can not be solved directly. The diameter of $\mathcal{B}_i$ is selected according to empirical rules in the LLE and ISOMAP which are local and global manifold learning algorithm respectively. In SMCE, this challenge is to be solved by a sparse optimization problem. Consider a sample point x_i and a sample set $\{x_j | j \neq i, j = 1, 2, \cdots, N\}$, the column vector c_i with dimension $N - 1$ is obtained by solved Eq. (2).

$$||[x_1 - x_i, \cdots, x_N - x_i]c_i||_2 \leqslant \epsilon \quad and \quad \mathbf{1}^T c_i = 1 \tag{2}$$

Where the solution c_i is sparse and the non-zero elements are corresponding to several sample points lying in the same manifold that x_i belongs to. It is the key difference between SMCE and other manifold learning algorithm.

In the case of densely sampled set, the affine subspace coincides with the d_l dimensional tangent space of $\mathcal{M}_l$ at x_i. In other words, the sample point corresponding to non-zeros elements of c_i may no longer be the closest points to x_i in $\mathcal{M}_l$. Therefore, the vectors $\{x_j - x_i\}_{j \neq i}$ is normalized and let

$$X_i := \left[\frac{x_1 - x_i}{||x_1 - x_i||_2}, \cdots, \frac{x_N - x_i}{||x_N - x_i||_2} \right] \in \mathbb{R}^{D \times (N-1)} \tag{3}$$

Thus the (2) has the following form:

$$||X_i c_i||_2 \leqslant \epsilon \quad and \quad \mathbf{1}^T c_i = 1 \tag{4}$$

In this way, among all the solutions of (4), the one that uses a few closest neighbors of x_i is searched by considering the following ℓ^1 optimization problem.

$$min \, ||Q_i c_i||_1 \quad s.t. \, ||X_i c_i||_2 \leqslant \epsilon, \quad \mathbf{1}^T c_i = 1 \tag{5}$$

Where the proximity inducing matrix Q_i, which is a positive definite diagonal matrix, is defined as

$$Q_i = \frac{||X_j - X_i||_2}{\sum_{t \neq i} ||X_t - X_i||_2} \in (0, 1] \tag{6}$$

Another optimization problem which is related to (5) by the method of Lagrange multipliers is

$$min \, \lambda ||Q_i c_i||_1 + \frac{1}{2} ||X_i c_i||_2^2 \quad s.t. \, \mathbf{1}^T c_i = 1 \tag{7}$$

Where the parameter λ sets the trade-off between the sparsity of the solution and the affine reconstruction error. By solving the optimization problem above for each sample point, it is obtained that the solution $c_i^T := [c_{i1}, \cdots, c_{iN}]$. Thus, the weight vector $w_i^T = [w_{i,1}, \cdots, w_{iN}] \in \mathbb{R}^N$ is defined as

$$w_{ii} = 0, \quad w_{ij} := \frac{c_{ij}/||X_j - X_i||_2}{\sum_{t \neq i} c_{it}/||X_t - X_i||_2}, \quad j \neq i \tag{8}$$

The similarity graph $G = (V, E)$ whose nodes represent sample points is built. Node i, corresponding to x_i, connects the node j, corresponding to x_j, with an edge whose weight is equal to $|w_{ij}|$. Each node i connects only a few other nodes named sparse neighbors that correspond to the neighbors of x_i in the same manifold. Hence, the similarity matrix W is constructed with weight vector w_i and reflect the distance from sparse neighbors to x_i. The samples are clustered by applying spectral clustering to W [9], or by applying dimensionality reduction for original samples with Laplacian eigenmap [10], and then the classifier is constructed effectively.

Table 1. Characters in time domain

k	y_k	k	y_k	k	y_k	k	y_k
1	Peak value	4	Root mean square value	7	Kurtosis	10	Impulsion index
2	Peak to peak value	5	Variance	8	Peak index	11	Wave index
3	Absolute mean value	6	Degree of skewness	9	Kurtosis index	12	Tolerance index

3   The Method of Feature Extraction Based on SMCE

The mathematical principles of SMCE adapt to engineering applications because the high-dimensional Euclidean space in SMCE is corresponding to the feature space that is spanned by feature vectors of machinery. In practice, the manifolds embedded in feature space have the following characters: the density of sample points in manifold is different in different regions in the manifold, and curvature of each point in manifold is also different. Therefore, it is inappropriate that the fixed neighborhood size is used to extract the manifold structure within mechanical operation data.

The initial feature space should be constructed firstly before that the non-linear dimensionality reduction is done by manifold learning. In the engineering application, the initial feature space can be spanned by feature vectors that should be composed by characters in time domain or characters in frequency domain, or characters in both time and frequency domain of vibration signal [1]. In this paper, we combine characters in time domain and frequency domain of vibration signals as an initial feature vector. The term of characters in time domain is shown in Table 1, and the sub-band energy method is employed to establish frequency feature vector.

The energy distribution in different frequency sub-bands of the vibration signals indicates the operation condition of machinery. Thus the sub-band energy vector could regard as the frequency feature vector. Divide vibration signal into m sub-bands with a constant bandwidth and the frequency range of i-th sub-band is shown in (9).

$$f_i = f_{i-1} + \Delta f \tag{9}$$

Where $i = 1, 2, \cdots, m$, f_i and f_{i-1} is the upper and low limiting frequency of i-th sub-band, Δf is the bandwidth of i-th sub-band. Through Fast Fourier Transform(FFT), the number of frequency lines n_i in each sub-band is equal, that is

$$n_i = \frac{f_i - f_{i-1}}{\xi} \tag{10}$$

Where ξ is frequency resolution. Hence, all frequency lines can be divided into m equal partitions related to m sub-bands. Suppose that the sequence index of frequency lines in i-th sub-band is corresponding to k_{i-1} and k_i, the energy of i-th sub-band is obtained by (11) according to Parseval's theorem.

$$E_i = \sum_{k=k_{i-1}}^{k_i} |A_k|^2 \Delta f \tag{11}$$

Where A_k is the amplitude of frequency line whose sequence index is k. In this paper, the vibration signal is divided into 16 equal sub-band, thus we can obtain a feature vector $\boldsymbol{y}_f$ of dimension 16. And $\boldsymbol{y}_f$ is normalized by (12). In a similar way, the feature vectors in time domain that is shown in Table 1 can be normalized, and we obtained a normalized feature vector in time domain Y_t with dimension 12. And then, the feature vectors $Y = [Y_f, Y_t]$ with dimension 28 is used to span the initial feature space.

$$Y_f = \frac{\boldsymbol{y}_f - min(\boldsymbol{y}_f)}{max(\boldsymbol{y}_f) - min(\boldsymbol{y}_f)} \tag{12}$$

The presented method of machinery fault diagnosis based on manifold learning is done with following stage.

(1) For each vibration signal, the feature vector with dimension 28 is constructed firstly.
(2) Using appropriate parameter λ to build similarity matrix W. Due to there is not mature theory to guide us to choose the best parameter λ, therefore the choice of best parameter λ is done by equal step length searching. Namely, the diagnosis accuracy is different under different λ, and we fixed the λ corresponding to best accuracy.
(3) Clustering of the samples data by applying spectral clustering [9] to W. The samples data from the same machinery condition is in one set and from the different condition is in different set. This is the unsupervised fault diagnosis approach. In addition, we apply laplacian eigenmap [10] to do dimensionality reduction and the low-dimensionality representation of original high-dimensionality feature space is obtained. And then, the support vector machine(SVM) is employed to do the classification of the low-dimensionality space. Because of SVM is one of supervised classifier, this approach of fault diagnosis is named supervised method of fault diagnosis.

4   Experiment

Bearing Date Center(BDC) [11] of Case Western Reserve University has been used by many researchers [12,13]. The availability of BDC has been proved, and this database has become a standard database of vibration signal of bearing defect. In this paper, The deep groove ball bearings SKF 6205-2RS-JEM-SKF with 9 rolling elements made by Swedish is employed as the analysis object. In experiment, the shaft speed is 1750RPM and the sampling frequency is 48KHZ. The vibration signal that collected by a accelerometer is shown in Fig. 1. Considering four condition of bearing, that is, normal, inner-race defect, ball defect, outer-race defect, 200 samples were chosen for each condition. Thus there are 800 samples in all and the sample length is 2048 of all. The feature space is spanned by 800 feature vectors, denoted by $\boldsymbol{x}_1, \boldsymbol{x}_2, \cdots, \boldsymbol{x}_{800} \in \mathbb{R}^{28}$, and the manifold structure in the feature space is extracted by the method described previously. Similarity matrix $W \in \mathbb{R}^{800 \times 800}$ is obtained with fixed parameter λ. And then,

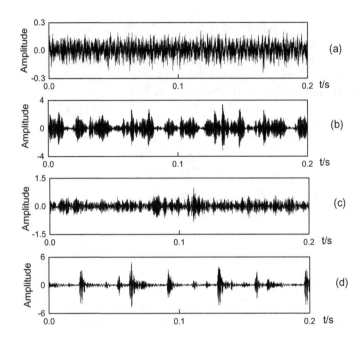

Fig. 1. The original signal of bearing. (a) Normal condition. (b) Inner-race defect. (c) Ball defect. (d) Outer-race defect.

the unsupervised fault diagnosis can be achieved by spectral clustering [9]. For the supervised fault diagnosis, Laplacian eigenmap [10] is applied to the W, and original feature space is mapped to three dimensional space. And then, the samples of each condition in three dimensional space is divided into 2 equal partition. One of the half samples are taken as training samples, and the remaining are testing samples. The problem we faced is to classify 4 conditions, but SVM can classify only 2 condition at a time. Generalizing method to solve this challenging is by decomposing the 4 class problem into 2 class problem. Each time take one class of the training samples as positive class and take remaining class as negative class, thus classification of 4 class is done by doing SVM 4 times. The parameter λ play a important role for diagnosis, we make the parameter λ to traverse [10,200] with step 5. Under different λ, the diagnosis accuracy of spectral clustering and support vector machine is shown in Fig. 2.

As the Fig. 2 show, the best accuracy of diagnosis with spectral clustering is up to 97.125 % when the λ is set as 30. In the other way, three dimensional representation of origin feature space with dimension 28 is obtained under different λ. The best accuracy of diagnosis with SVM is up to 98.75 % when the λ is set as 130. And the three dimensional space is shown in Fig. 3. Spectral clustering do the classification using the geometric structure of space spanned by sample point and without the tag of samples. This is the reason that it is named unsupervised diagnosis. Conversely, SVM do the classification with tags of samples,

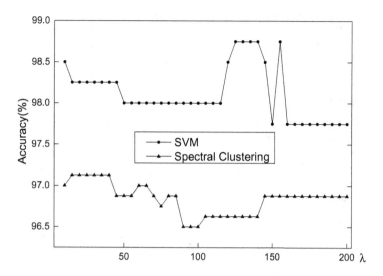

Fig. 2. The diagnosis accuracy with different λ

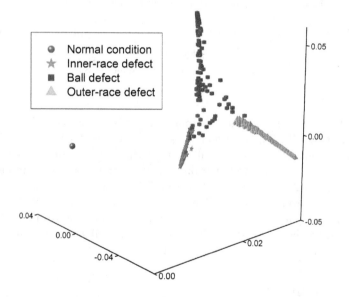

Fig. 3. Three dimensional representation of original feature space

thus the diagnosis based on SVM is named supervised diagnosis. From Fig. 3, the diagnosis accuracy of SVM is higher than spectral clustering and the accuracy of both has remained stable under different λ. The accuracy difference between maximum and minimum is 0.625 % and 1 % with spectral clustering and SVM respectively. Namely, the parameter λ is easily fixed. It is demonstrated that the presented method is suitable for engineering applications. To reveal the out-

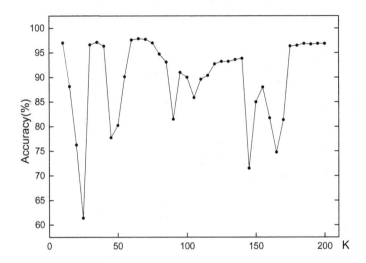

Fig. 4. The diagnosis accuracy with LLE and SVM

Table 2. The comparison of accuracy between SMCE and LLE

Diagnosis method	Minimum	Maximum	Mean	Standard variance
SMCE	97.75 %	98.75 %	98.12 %	0.32 %
LLE	61.38 %	97.88 %	89.66 %	8.57 %

standing characters of presented method, the fault diagnosis is done with LLE [5] instead of SMCE. The classifier is established with SVM. The accuracy of diagnosis is shown in Fig. 4. It is illustrated that the diagnosis accuracy vary in a large range with different parameter. The comparison of diagnosis accuracy between SMCE and LLE is shown in Table 2. The diagnosis method with SMCE is more stable than LLE. The satisfactory result is reached with nearly almost arbitrary parameter with SMCE, but the diagnosis result with LLE is more affected by the parameters. Therefore the convenience of parameter choice is one of the advantages of presented method.

5   Conclusion

(1) The method of feature extraction based on SMCE could extract low dimensional manifold structure that indicates the nature of mechanical condition embed in high-dimensional feature space.
(2) The feature vector that consist of time-domain characters and sub-band energy can be used to diagnose the condition of machinery.
(3) The diagnosis accuracy of presented method in this paper is less affected by the parameters, therefore it is more suitable for engineering applications thanks to the convenience of parameter choice.

References

1. Wang, J.: The Fault Diagnosis Technology of Mechanical Equipment and its Application. Northwestern Polytechnical University Press, Xi'an (2001)
2. Wold, S., Esbensen, K., Geladi, P.: Principal component analysis. Chemometr. Intell. Lab. Syst. **2**(1–3), 37–52 (1987)
3. Chiang, L.H., Kotanchek, M.E., Kordon, A.K.: Fault diagnosis based on fisher discriminant analysis and support vector machines. Comput. Chem. Eng. **28**(8), 1389–1401 (2004)
4. Sebastian Seung, H.: The manifold ways of perception. Science **290**, 2268–2269 (2000)
5. Roweis, S.T., Saul, L.K.: Nonlinear dimensionality reduction by locally linear embedding. Science **290**(5500), 2323–2326 (2000)
6. Tenenbaum, J.B., De Silva, V., Langford, J.C.: A global geometric framework for nonlinear dimensionality reduction. Science **290**(5500), 2319–2323 (2000)
7. Jiang, Q., Jia, M., Jianzhong, H., Feiyun, X.: Machinery fault diagnosis using supervised manifold learning. Mech. Syst. Sig. Proc. **23**(7), 2301–2311 (2009)
8. Elhamifar, E., Vidal, R.: Sparse manifold clustering and embedding. In: Advances in Neural Information Processing Systems, pp. 55–63 (2011)
9. Ng, A.Y., Jordan, M.I., Weiss, Y.: On spectral clustering: analysis and an algorithm. Adv. Neural Inf. Process. Syst. **14**, 849–856 (2002)
10. Belkin, M., Niyogi, P.: Laplacian eigenmaps and spectral techniques for embedding and clustering. Adv. Neural Inf. Process. Syst. **14**(6), 585–591 (2002)
11. Loparo, K.A.: Bearing vibration data set. http://csegroups.case.edu/bearingdatacenter/home
12. Zhang, L., Jinwu, X., Yang, J., Yang, D., Wang, D.: Multiscale morphology analysis and its application to fault diagnosis. Mech. Syst. Sig. Proc. **22**(3), 597–610 (2008)
13. Yang, H., Mathew, J., Ma, L.: Fault diagnosis of rolling element bearings using basis pursuit. Mech. Syst. Sig. Proc. **19**(2), 341–356 (2005)

p-Spectral Clustering Based on Neighborhood Attribute Granulation

Shifei Ding[1,2(✉)], Hongjie Jia[1,2], Mingjing Du[1,2], and Qiankun Hu[1,2]

[1] School of Computer Science and Technology, China University of Mining and Technology, Xuzhou 221116, China
{dingsf,jiahongjie}@cumt.edu.cn

[2] Key Laboratory of Intelligent Information Processing, Institute of Computing Technology, Chinese Academy of Sciences, Beijing 100190, China

Abstract. Clustering analysis is an important method for data mining and information statistics. Data clustering is to find the intrinsic links between objects and describe the internal structures of data sets. *p*-Spectral clustering is based on Cheeger cut criterion. It has good performance on many challenging data sets. But the original *p*-spectral clustering algorithm is not suitable for high-dimensional data. To solve this problem, this paper improves *p*-spectral clustering using neighborhood attribute granulation and proposes NAG-pSC algorithm. Neighborhood rough sets can directly process the continuous data. We introduce information entropy into the neighborhood rough sets to weaken the negative impact of noise data and redundant attributes on clustering. In this way, the data points within the same cluster are more compact, while the data points between different clusters are more separate. The effectiveness of the proposed NAG-pSC algorithm is tested on several benchmark data sets. Experiments show that the neighborhood attribute granulation will highlight the differences between data points while maintaining their characteristics in the clustering. With the help of neighborhood attribute granulation, NAG-pSC is able to recognize more complex data structures and has strong robustness to the noise or irrelevant features in high-dimensional data.

Keywords: *p*-Spectral clustering · Rough set · Attribute reduction · Neighborhood granulation

1 Introduction

Spectral clustering treats clustering problem as a graph partitioning problem. It can solve the graph cut objective function using the eigenvectors of graph Laplacian matrix [1]. Compared with the conventional clustering algorithms, spectral clustering is able to recognize more complex data structures, especially suitable for non-convex data sets. Recently, an improved version of normalized cut named Cheeger cut has aroused much attention [2]. Research shows that Cheeger cut is able to produce more balanced clusters through graph *p*-Laplacian matrix [3]. *p*-Laplacian matrix is a nonlinear generalization form of graph Laplacian.

p-spectral clustering is based on Cheeger cut to group data points. As it has solid theoretical foundation and good clustering results, the research in this area is very

Z. Shi et al. (Eds.): IIP 2016, IFIP AICT 486, pp. 50–58, 2016.
DOI: 10.1007/978-3-319-48390-0_6

active at present. Dhanjal et al. present an incremental spectral clustering which updates the eigenvectors of the Laplacian in a computationally efficient way [4]. Gao et al. construct the sparse affinity graph on a small representative dataset and use local interpolation to improve the extension of the clustering results [5]. Semertzidis et al. inject the pairwise constraints to a small affinity sub-matrix and use a sparse coding strategy of a landmark spectral clustering to preserve low complexity [6].

Nowadays, science and technology is growing by leaps and bounds and massive data result in "data explosion". These data are often accompanied by high dimensions. When dealing with high-dimensional data, some clustering algorithms that perform well in low-dimensional data space are often unable to get good clustering results, and even invalid [7]. Attribute reduction is an effective way to decrease the size of data, and it is often used as a preprocessing step for data mining. The essence of attribute reduction is to remove irrelevant or unnecessary attributes while maintaining the classification ability of knowledge base. Efficient attribute reduction not only can improve the knowledge clarity in intelligent information systems, but also reduce the cost of information systems to some extent. In order to effectively deal with high-dimensional data, we design a novel attribute reduction method based on neighborhood granulation and combine it with *p*-spectral clustering. The proposed algorithm inherits the advantages of neighborhood rough set and graph *p*-Laplacian. Its effectiveness is demonstrated by comprehensive experiments on benchmark data sets.

This paper is organized as follows: Sect. 2 introduces *p*-spectral clustering; Sect. 3 uses information entropy to improve the attribute reduction based on neighborhood rough sets; Sect. 4 improves *p*-spectral clustering with the neighborhood attribute granulation; Sect. 5 verifies the effectiveness of the proposed algorithm using benchmark data sets; finally, we summarize the main contribution of this paper.

2 *p*-Spectral Clustering

The idea of spectral clustering comes from spectral graph partition theory. Given a data set, we can construct an undirected weighted graph $G = (V, E)$, where V is the set of vertices represented by data points, E is the set of edges weighted by the similarities between the edge's two vertices. Suppose A is a subset of V, the complement of A is written as $\bar{A} = V \backslash A$. The cut of A and $\bar{A}$ is defined as:

$$cut(A, \bar{A}) = \sum_{i \in A, j \in \bar{A}} w_{ij} \tag{1}$$

where w_{ij} is the similarity between vertex i and vertex j.

In order to get more balanced clusters, Cheeger et al. propose Cheeger cut criterion, denoted as *Ccut* [8]:

$$Ccut(A, \bar{A}) = \frac{cut(A, \bar{A})}{\min\{|A|, |\bar{A}|\}} \tag{2}$$

where $|A|$ is the number of data points in set A. Cheeger cut is to minimize formula (2) to get a graph partition. The optimal graph partition means that the similarities within a cluster are as large as possible, while the similarities between clusters are as small as possible. But according to the Rayleigh quotient principle, calculating the optimal Cheeger cut is an NP-hard problem. Next we will try to get an approximate solution of Cheeger cut by introducing p-Laplacian into spectral clustering.

Hein et al. define the inner product form of graph p-Laplacian Δ_p as follows [9]:

$$\langle f, \Delta_p f \rangle = \frac{1}{2} \sum_{i,j=1}^{n} w_{ij} (f_i - f_j)^p \tag{3}$$

where $p \in (1,2]$, f is the eigenvector of p-Laplacian matrix.

Theorem 1. For $p > 1$ and every partition of V into $A, \bar{A}$ there exists a function (f, A) such that the functional F_p associated to the p-Laplacian satisfies

$$F_p(f,A) = \frac{\langle f, \Delta_p f \rangle}{\|f\|^p} = cut(A,\bar{A}) \left| \frac{1}{|A|^{\frac{1}{p-1}}} + \frac{1}{|\bar{A}|^{\frac{1}{p-1}}} \right|^{p-1} \tag{4}$$

where $\|f\|^p = \sum_{i=1}^{n} |f_i|^p$. The expression (4) can be interpreted as a balanced graph cut criterion, and we have the special cases

$$\lim_{p \to 1} F_p(f,A) = Ccut(A,\bar{A}) \tag{5}$$

Theorem 1 shows that Cheeger cut can be solved in polynomial time using p-Laplacian operator. So the solution of $F_p(f)$ is a relaxed approximate solution of Cheeger cut and the optimal solution can be obtained by the eigen-decomposition of p-Laplacian:

$$\lambda_p = \arg \min_{p \to 1} F_p(f) \tag{6}$$

where λ_p is the eigenvalue corresponding to eigenvector f.

Specifically, the second eigenvector $v_p^{(2)}$ of p-Laplacian matrix will lead to a bipartition of the graph by setting an appropriate threshold [3]. The optimal threshold is determined by minimizing the corresponding Cheeger cut. For the second eigenvector $v_p^{(2)}$ of graph p-Laplacian Δ_p, the threshold should satisfy:

$$\arg \min_{A_t = \{i \in V | v_p^{(2)}(i) > t\}} Ccut(A_t, \bar{A}_t) \tag{7}$$

3   Neighborhood Attribute Granulation

Rough set theory is proposed by professor Pawlak in 1982 [10]. Attribute reduction is one of the core contents of rough set knowledge discovery. However, Pawlak rough set is only suitable for discrete data. To solve this problem, Hu et al. propose neighborhood rough set model [11]. This model can directly analyze the attributes with continuous values. Therefore, it has great advantages in feature selection and classification accuracy.

Definition 1. Domain $U = \{x_1, x_2, \cdots, x_n\}$ is a non-empty finite set in real space, for $x_i \in U$, the δ-neighborhood of x_i is defined as:

$$\delta(x_i) = \{x | x \in U, \Delta(x, x_i) \le \delta\} \tag{8}$$

where $\delta \ge 0$, $\delta(x_i)$ is called the neighborhood particle of x_i, Δ is a distance function.

Definition 2. Given a domain $U = \{x_1, x_2, \cdots, x_n\}$ located in real space. A represents the attribute set of U; D represents the decision attribute. If A is able to generate a family of neighborhood relationship of domain U, then $NDT = \langle U, A, D \rangle$ is called a neighborhood decision system.

For a neighborhood decision system $NDT = \langle U, A, D \rangle$, domain U is divided into N equivalence classes by decision attribute $D : X_1, X_2, \cdots, X_N$. $\forall B \subseteq A$, the lower approximation is $\underline{N_B}D = \bigcup_{i=1}^{N} \underline{N_B}X_i$, where $\underline{N_B}X_i = \{x_i | \delta_B(x_i) \subseteq X_i, x_i \in U\}$.

According to the nature of lower approximation, we can define the dependence of decision attribute D on condition attribute B:

$$\gamma_B(D) = \frac{Card(\underline{N_B}D)}{Card(U)} \tag{9}$$

where $0 \le \gamma_B(D) \le 1$. Obviously, the greater the positive region $\underline{N_B}D$, the stronger the dependence of decision D on condition B.

Definition 3. Given a neighborhood decision system $NDT = \langle U, A, D \rangle$, $B \subseteq A$, $\forall a \in A - B$, then the significant degree of a relative to B is defined as:

$$SIG(a, B, D) = \gamma_{B \cup a}(D) - \gamma_B(D) \tag{10}$$

However, sometimes several attributes may have the same greatest importance degree. Traditional reduction algorithms take the approach of randomly choosing one of the attributes, which is obviously arbitrary does not taking into account the impact of other factors on attribute selection and may lead to poor reduction results. From the viewpoint of information theory to analyze attribute reduction can improve the reduction accuracy [12]. Here, we use information entropy as another criterion to evaluate attributes. The definition of entropy is given below.

Definition 4. Given knowledge P and its partition $U/P = \{X_1, X_2, \cdots, X_n\}$ exported on domain U. The information entropy of knowledge P is defined as:

$$H(P) = -\sum_{i=1}^{n} p(X_i) \log p(X_i) \tag{11}$$

where $p(X_i) = |X_i|/|U|$ represents the probability of equivalence class X_i on domain U.

If multiple attributes have the same greatest importance degree, then we may compare their information entropy and select the attribute with the minimum entropy (because it carries the least uncertain information). Incorporate the selected attribute into the reduction set, and repeat this process for each attribute until the reduction set no longer changes. This improved attribute reduction algorithm is shown as Algorithm 1.

Algorithm 1. Neighborhood attribute granulation with information entropy.

Input: Neighborhood decision system $NDT = \langle U, A, D \rangle$.

Output: The reduced attribute set red.

Step 1. $\forall a \in A$, calculate the neighborhood relation N_a.

Step 2. Initialize $red = \varnothing$.

Step 3. $\forall a_i \in A - red$, calculate the importance degree of each attribute $SIG(a_i, red, D) = \gamma_{red \cup a}(D) - \gamma_{red}(D)$.

Step 4. If $\max_i(SIG(a_i, red, D))$ contains only one attribute, then select a_k so as to satisfy $SIG(a_k, red, D) = \max_i(SIG(a_i, red, D))$; otherwise, calculate the information entropy of these attributes and select a_k so as to satisfy $H(a_k) = \min_i(H(a_i))$.

Step 5. If $SIG(a_k, red, D) > 0$, put a_k into the reduction set, $red = red \cup a_k$, then go to step 3; otherwise, output red and the algorithm ends.

4 p-Spectral Clustering Based on Neighborhood Attribute Granulation

Massive high-dimensional data processing has been a challenge problem in data mining. High-dimensional data is often accompanied by the "curse of dimensionality", so traditional p-spectral clustering algorithms cannot play to their strengths very well. Moreover, real data sets often contain noise and irrelevant features, likely to cause "dimension trap". It would interfere with the clustering process of algorithms, affecting the accuracy of clustering results [13]. To solve this problem, we propose a novel p-spectral clustering algorithm based on neighborhood attribute granulation (NAG-pSC). The detailed steps of NAG-pSC algorithm is given in Algorithm 2.

Algorithm 2. *p*-Spectral clustering algorithm based on neighborhood attribute granulation.

Input: dataset $X = \{x_1, x_2, \ldots, x_n\}$, the cluster number k.

Output: k divided clusters.

Step 1. Reduce the attributes of data points according to Algorithm 1 and obtain the reduced attribute set *red*.

Step 2. Affter attribute granulation, calculate the similarities between data points based on the new data set *red*, and form the affinity matrix $W \in \mathbb{R}^{n \times n}$ using Gaussian kernel.

Step 3. Initialize the first cluster $A_1 = V$ and set the cluster number $s = 1$.

Step 4. Repeat from Step 4 to Step 8.

Step 5. Construct *p*-Laplacian matrix according to formula [3] with the affinity matrix W.

Step 6. Calculate the second eigenvector $v_p^{(2)}$ of graph *p*-Laplacian Δ_p, and search an appropriate threshold value that satisfies formula [7].

Step 7. Use $v_p^{(2)}$ to split each cluster $A_i (i = 1, 2, \ldots, s)$ and minimize the overall Cheeger cut objective function.

Step 8. $s \Leftarrow s + 1$.

Step 9. When the number of clusters $s == k$, stop the loop and output the clustering results.

5   Experimental Analysis

To test the effectiveness of the proposed NAG-pSC algorithm, we use six benchmark data sets to do the experiments. The characteristics of these data sets are shown in Table 1.

Table 1. Data sets used in the experiments

Data set	Instance number	Condition attribute number	Decision attribute number	Class number
Sonar	208	60	1	2
WDBC	569	30	1	2
Colon Cancer	62	2000	1	2
Duke Breast Cancer	44	7129	1	2

In this paper, we use F-measure to evaluate the merits of clustering results [14]. The F-score of each class i and the total F index of the clustering results are defined as:

$$F(i) = \frac{2 \times P(i) \times R(i)}{P(i) + R(i)} \tag{12}$$

$$F = \frac{1}{n} \sum_{i=1}^{k} [N_i \times F(i)] \tag{13}$$

where $P(i) = N_{ii*}/N_{i*}$ is the precision rate and $R(i) = N_{ii*}/N_i$ is the recall rate; N_{ii*} is the size of the intersection of class i and cluster i^*; N_i is the size of class i; N_{i*} is the size of cluster i^*; n is the number of data points; k is the class number; N_i is the size of class i. $F \in [0, 1]$, the greater the F index is, means the clustering results of the algorithm is closer to the real data category.

In the experiment, NAG-pSC algorithm is compared with the traditional spectral clustering (SC), density sensitive spectral clustering (D-SC) [1] and p-spectral clustering (pSC) [3]. The threshold δ is important in neighborhood rough set. Hu et al. recommend a value range [0.2, 0.4] of δ based on experimental analysis [11]. So we set the neighborhood size δ via a cross-validatory search in the range [0.2, 0.4] (with step size 0.05) for each data set. The clustering results of these four algorithms are shown in Fig. 1. The horizontal axis of the figure is the cluster label, and the vertical axis is the F-score of each cluster.

From Fig. 1 we can see that, the performance of SC algorithm is close to D-SC algorithm. This is mainly because that they all based on graph theory and turn the clustering problem into a graph partitioning problem. Using the p-Laplacian transform, pSC may find the global optimum solution. SC works well on Sonar data set. D-SC deals well with Colon Cancer data set. pSC can generate balanced clusters on WDBC

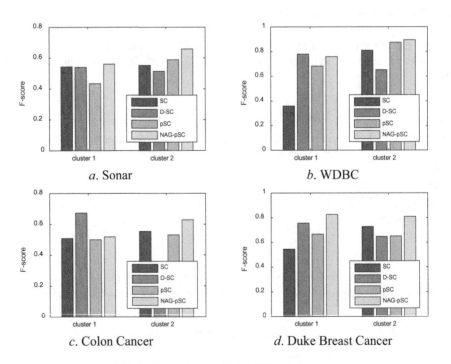

a. Sonar   *b.* WDBC

c. Colon Cancer   *d.* Duke Breast Cancer

Fig. 1. Clustering results on different datasets

data set. But for high dimensional clustering problems, their F-scores are lower than the proposed NAG-pSC algorithm. Because the information in each attribute of the instances is different, and they also make different contributions to the clustering. Improper feature selection would cause a greate impact on the clustering results. Traditional spectral clustering algorithm does not take this into account, susceptible to the interference of noise and irrelevant attributes. For further comparison, Table 2 lists the overall *F* index for each algorithm and the number of condition attributes of different data sets.

Table 2. Total *F* index of different algorithms

Data set		Algorithm			
		SC	D-SC	pSC	NAG-pSC
Sonar	*F* index	0.5483	0.5314	0.5170	0.6126
	Condition attributes	60	60	60	17
WDBC	*F* index	0.6408	0.7331	0.8019	0.8443
	Condition attributes	30	30	30	13
Colon Cancer	*F* index	0.5377	0.5453	0.5202	0.5895
	Condition attributes	2000	2000	2000	124
Duke Breast Cancer	*F* index	0.6322	0.7177	0.6593	0.8182
	Condition attributes	7129	7129	7129	167

Table 2 shows that NAG-pSC algorithm can well deal with high-dimensional data. NAG-pSC algorithm uses neighborhood rough sets to optimize data instances. The neighborhood attribute reduction based on information entropy diminishes the negative impact of noise data and redundant attributes on the clustering. So in most cases, NAG-pSC algorithm has higher clustering accuracy. NAG-pSC algorithm combines the advantages of *p*-spectral clustering and neighborhood attribute granulation. It has good robustness and strong generalization ability.

6 Conclusions

To improve the performance of *p*-spectral clustering on high-dimensional data, we modify the attribute reduction method based on neighborhood rough sets. In the new method, the attribute importance is combined with information entropy to select the appropriate attributes. Then we propose NAG-pSC algorithm based on the optimized attribute reduction set. Experiments show that NAG-pSC algorithm is superior to traditional spectral clustering, density sensitive spectral clustering and *p*-spectral clustering. In the future, we will study how to apply NAG-pSC algorithm to web data mining, image retrieval and other realistic scenes.

Acknowledgements. This work is supported by the National Natural Science Foundation of China (Nos. 61379101, 61672522), and the National Key Basic Research Program of China (No. 2013CB329502).

References

1. Yang, P., Zhu, Q., Huang, B.: Spectral clustering with density sensitive similarity function. Knowl.-Based Syst. **24**(5), 621–628 (2011)
2. Bresson, X., Szlam, A.D.: Total variation and cheeger cuts. In: Proceedings of the 27th International Conference on Machine Learning, pp. 1039–1046 (2010)
3. Bühler, T., Hein, M.: Spectral clustering based on the graph p-Laplacian. In: Proceedings of the 26th International Conference on Machine Learning, pp. 81–88 (2009)
4. Dhanjal, C., Gaudel, R., Clémençon, S.: Efficient eigen-updating for spectral graph clustering. Neurocomputing **131**, 440–452 (2014)
5. Cao, J., Chen, P., Dai, Q., et al.: Local information-based fast approximate spectral clustering. Pattern Recogn. Lett. **38**, 63–69 (2014)
6. Semertzidis, T., Rafailidis, D., Strintzis, M.G., et al.: Large-scale spectral clustering based on pairwise constraints. Inf. Process. Manage. **51**, 616–624 (2015)
7. Jia, H., Ding, S., Xu, X., et al.: The latest research progress on spectral clustering. Neural Comput. Appl. **24**(7–8), 1477–1486 (2014)
8. Jia, H., Ding, S., Du, M.: Self-tuning p-Spectral clustering based on shared nearest neighbors. Cogn. Comput. **7**(5), 622–632 (2015)
9. Hein, M., Audibert, J.Y., Von Luxburg, U.: Graph Laplacians and their convergence on random neighborhood graphs. J. Mach. Learn. Res. **8**(12), 1325–1368 (2007)
10. Pawlak, Z.: Rough sets. Int. J. Comput. Inform. Sci. **11**(5), 341–356 (1982)
11. Hu, Q.H., Yu, D.R., Xie, Z.X.: Numerical attribute reduction based on neighborhood granulation and rough approximation. J. Softw. **19**(3), 640–649 (2008)
12. Ding, S.F., Zhu, H., Xu, X.Z., et al.: Entropy-based fuzzy information measures. Chin. J. Comput. **35**(4), 796–801 (2012)
13. Du, M., Ding, S., Jia, H.: Study on density peaks clustering based on k-nearest neighbors and principal component analysis. Knowl.-Based Syst. **99**, 135–145 (2016)
14. Jia, H., Ding, S., Zhu, H., et al.: A feature weighted spectral clustering algorithm based on knowledge entropy. J. Softw. **8**(5), 1101–1108 (2013)

Assembly Sequence Planning Based on Hybrid Artificial Bee Colony Algorithm

Wenbing Yuan, Liang Chang[(✉)], Manli Zhu, and Tianlong Gu

Guangxi Key Laboratory of Trusted Software,
Guilin University of Electronic Technology, Guilin 541004, China
changl@guet.edu.cn

Abstract. Intelligent algorithm provides a promising approach for solving the Assembly Sequence Planning (ASP) problem on complex products, but there is still challenge in finding best solutions efficiently. In this paper, the artificial bee colony algorithm is modified to deal with this challenge. The algorithm is modified from four aspects. First, for the phase that employed bee works, a simulated annealing operator is introduced to enrich the diversity of nectar sources and to enhance the local searching ability. Secondly, in order to prevent the swarm from falling into local optimal solutions quickly, a tournament selection mechanism is introduced for the onlooker bees to choose the food source. Thirdly, for the phase that scout bee works, a learning mechanism is introduced to improve the quality of new generated food sources and to increase the convergence speed of the algorithm. Finally, a fitness function based on the evaluation indexes of assemblies is proposed to evaluate and select nectar sources. The experimental results show that the modified algorithm is effective and efficient for the ASP problem.

Keywords: Artificial bee colony algorithm · Simulated annealing operators · Assembly sequence planning · Multi-objective optimization

1 Introduction

Assembly sequence planning (ASP) aims to find a proper sequence of assembly operations under some operational constraints and precedence constraints. ASP is an important manufacturing process in that the quality of assembly has a direct effect on the performance of product. According to a statistical report, a good assembly sequence can reduce costs of manufacturing about 20 %–40 %, and can increase the productivity about 100 %–200 %.

Traditional methods of assembly sequence planning are based on cut-sets [1] or assembly knowledge representation [2]. A shortcoming of these methods is that they have to face the combinatorial state explosion problem caused by the increasing of the number of parts [3]. During the past 20 years, many researchers tried to apply intelligent algorithms to solve the ASP problem [4–9] and got many good results.

Artificial bee colony (ABC) algorithm is an artificial intelligence algorithm that was proposed for the multi-variable and multi-modal optimization of continuous functions. It simulates the emergent intelligent behaviour of foraging bees in three phases. At the

© IFIP International Federation for Information Processing 2016
Published by Springer International Publishing AG 2016. All Rights Reserved
Z. Shi et al. (Eds.): IIP 2016, IFIP AICT 486, pp. 59–71, 2016.
DOI: 10.1007/978-3-319-48390-0_7

initial phase, the scout bees start to explore the environment randomly in order to find a nectar source. After finding a nectar source, a bee becomes an employed bee and starts to exploit the discovered source; it returns to the hive with the nectar and share information about the source site by performing a dance on the dance area. Onlooker bees waiting in the hive watch the dances and choose a nectar source through the information shared by employed bees by using a wheel selection. If a source is exhausted, then the employed bees that are working in that source become scouts again and start to randomly search for new sources. Compared with the genetic algorithm [4], the firely algorithm [6] and the particle swarm optimization algorithm [7], the ABC algorithm is more competitive in the ability of global searching and the speed of convergence. It has been used to solve the global optimization problem [10], global numerical optimization [11], real-parameter optimization [12], shop scheduling problem [13] and disassembly planning problem [14].

In this paper, we apply the ABC algorithm to solve the assembly sequence planning problem. In order to improve the ability of searching optimal solution; we embed the simulated annealing algorithm and a local searching algorithm into the ABC algorithm. At the same time, we use a tournament selection mechanism to replace the wheel selection mechanism for escaping from local optimum. Furthermore, we apply a cooperative learning mechanism in the scout bee phase to improve the convergence speed, and propose a suitable fitness function to evaluate and filter nectar sources. In the rest of this paper, we call the improved algorithm as hybrid artificial bee colony algorithm (HABC).

2 Optimization Model of Assembly Sequences

In this paper, we use the following assumptions to simplify the optimization model of assembly sequences. (1) All the parts are rigid. In another word, no part is deformable in the assembly process. (2) The assembly directions are restricted to $\pm X, \pm Y, \pm Z$ direction in the three orthogonal coordinate axes. (3) Along each direction only one part is assembled, and in each assembly operation process only one tool is used.

An evaluation index for the fitness function is proposed to evaluate and filter the nectar sources: geometric feasibility, assembly stability, the number of parts violating local assembly precedence, and changes of assembly directions, tools and connections. The fitness function is applied to find the optimal nectar sources which are the optimal or near-optimal assembly sequences.

2.1 Geometric Feasibility

An assembly sequence satisfying geometric feasibility can ensure that the movement path of part is collision-free. In this paper, we use Cartesian 6 coordinate to represent the assembly directions $d_k = \{\pm x, \pm y, \pm z\}$. The values of 0 and 1 represent the interference between the parts. The I_{ijd_k} represents the interference in the d_k direction between p_i and p_j, which is shown in formula (1);

$$I_{ijd_k} = \begin{cases} 0, \text{ there is no interference in the } d_k \text{ direction between } p_i \text{ and } p_j \\ 1, \text{ there is an interference in the } d_k \text{ direction between } p_i \text{ and } p_j \end{cases} \quad (1)$$

We assume an assembly sequence $AS = \{p_1, p_2, \ldots, p_n\}$, when the subsequence $AS_1 = \{p_1, p_2, \ldots, p_m\}$ was assembled and p_i will be assembled, we represent the sum total of interference between p_i and all parts of AS_1 in the d_k direction with $S_k(p_i)(k = 1, 2, \ldots, 6)$ which is shown in formula (2):

$$S_k(p_i) = \sum_{j=1}^{m} I_{p_i p_j d_k} \quad (2)$$

$S_k(p_i) = 0$ indicates that the part p_i can be assembled in the d_k direction; otherwise it cannot. Then we can get a set of assembly directions of the part p_i: $DC(p_i) = \{d_k | S_k(p_i) = 0\}$. For any part p_i, if $DC(p_i) \neq \emptyset$, AS is a geometric feasible assembly sequence, infeasible otherwise. We represent the number of interference of AS with n_f, that is the number of $DC(p_i) = \emptyset$.

2.2 Assembly Stability

Parts or subassembly may be unstable, for its own gravity in the actual assembly process. Auxiliary devices are needed to maintain the stability of parts in the assembly process once some parts are unstable. For this reason, we consider the assembly stability as an evaluation index.

The tight fit, interference fit, screw, rivet and welding are the strong assembly connections which are the stable connections. The clearance fit, sticking and attachment are the weak connections; the surface mating and other fits without force are the unstable connections [15]. The values of s_{ij} represent the assembly connections between the p_i and p_j: $s_{ij} = 2$ indicates that they are joined by strong connections; $s_{ij} = 1$ indicates that they are joined by weak connections; $s_{ij} = 0$ indicates that they are joined by surface mating without force. Then the assembly stability can be represented with the stability matrix which is shown in formula (3):

$$SM = \left[s_{ij} \right]_{n \times n} \quad (3)$$

It is a stable connection if $s_{ij} = 2$, otherwise not.

Given that the temporary stable subassembly is represented as $AS_1 = \{p_1, p_2, \ldots, p_m\}$ (m is the number of parts which have been assembled), the part p_i is to be assembled in the next step. The assembly stability of the part p_i with AS_1 can be concluded depending on the stability matrix SM. The correlative stability value $s_{ij}(1 \leq j \leq m)$ can be inferred as $\{s_{i1}, s_{i2}, \ldots, s_{im}\}$. If $s_{ij} = 2(1 \leq j \leq m)$, the part p_i and AS_1 will be stable after they are assembled. Otherwise, the part p_i needs auxiliary devices in the assembly process. The assembly stability of each part will be checked and the number of parts n_s that make the subassembly self-stable in the assembly sequence can be computed.

2.3   The Number of Parts Violating Local Assembly Precedence

The reasonable assembly sequences not only satisfy the geometrical feasibility, but also the constraint relationships between parts. Therefore, the assembly precedence relationships of parts must be taken into consideration. For the complex assemblies, there will be too many possible assembly sequences, but most of them are actually not ensured to be correct. Even if the assembly sequences satisfy the geometrical feasibility, some parts maybe violate the local assembly precedence still. So the violating assembly precedence is another considerable index to evaluate assembly sequences. To an assembly sequence $AS = \{p_1, p_2, \ldots, p_n\}$, the number of parts violating local assembly precedence n_p can be obtained easily by assembly precedence matrix (APM) described in Sect. 3.2.

2.4   Changes of Assembly Directions, Tools And connections

The changes of the assembly directions, the assembly tools and the assembly connection types in the assembly process have a great effect on the assembly efficiency and cost. To improve the assembly efficiency and reduce the assembly cost, the parts assembled in the assembly direction, and the assembly connection types and assembly tools are suggested to be assembled in groups. The assembly directions of parts can be derived from the assembly interference matrix. Once the assembly sequence is confirmed, the number of change n_d of the assembly directions of all parts can be counted.

The assembly tools used to assemble one part may be multiple. The selected tools to assemble the current part should be the same as those used to assemble the former part so that the time for changing the tools can be shortened. All the assembly tools can be represented by a tool matrix $TM = [t_{ij}]_{n \times m}$, here n is the number of parts and m is the number of practicable tools to assemble the corresponding part. After the optimal or near-optimal assembly sequences have been generated, the corresponding tools are also confirmed at the same time and the number of change n_t of the assembly tools can be obtained.

It is assumed that the connection type for assembling each part is unique and the connection types of parts are determined in the design process. The assembly connections to assemble the parts can be represented as a vector of connection $CM = [c_{ij}]_{n \times 1}$. When the assembly sequences have been ascertained, the number of change n_t of the assembly connection types can also be acquired.

2.5   Fitness Function

Each evaluation index has different effects in the assembly process, so it is reasonable to put different weights on each index. Then the fitness function can be defined as the formula (4) and (5) by using the different weights:

$$f_1 = c_s n_s + c_p n_p + c_d n_d + c_t n_t + c_c n_c \tag{4}$$

$$f_2 = c_f n_f \tag{5}$$

c_f, c_s, c_p, c_d, c_t, and c_c are the weights of the evaluation indices respectively.

Fitness function is the metric of evaluating assembly sequences. Similarly, it is the metric of evaluating nectar sources in the search process. The smaller the value, the better the nectar source quality, and the assembly sequence is better correspondingly. If an assembly sequence satisfies the geometrical feasibility, formula (4) will be used to calculate the assembly cost; otherwise formula (4) and (5) will be used.

3   Assembly Sequences of HABC

The basic ABC algorithm was first used to solve optimization problems of continuous functions. A hybrid ABC algorithm is proposed to solve multi-objective optimization problems in the assembly sequence planning. HABC algorithm makes full use of swarm intelligence sharing information to accelerate search efficiency.

3.1   Nectar Source Code

In HABC algorithm, nectar position can be represented with the decimal sequence. Each nectar source is an assembly sequence. Profitability of nectar sources is used to measure the quality of assembly sequences. Each number and its location in the sequence represent the number and the assembly order of parts respectively. Suppose that a product is made up of 9 parts, the coding scheme can be described in Fig. 1:

Assembly sequence:	3	6	8	1	4	9	5	2	7
Sequence order:	1	2	3	4	5	6	7	8	9

Fig. 1.   Nectar source code

If the assembly sequence of Fig. 1 is a feasible sequence, the parts are assembled orderly from 3 to 7. Its quality is evaluated by the fitness function.

3.2   Initial Nectar Source

If the initial nectar sources are produced random completely, it will generate large amount of infeasible sequences which can affect the efficiency of algorithm. If all initial nectar sources are feasible sequences, it will affect the diversity of nectar sources. To guarantee an initial population with certain quality and diversity, a portion of nectar sources are generated by using some priority rules whereas the rest are produced randomly.

The assembly precedence matrix (APM) is used to generate partial sequences as the initialization nectar sources in a generator (Fig. 5) assembly [16]. A value of $b_{ij} = 1$ represents that part c_i must be assembled before part c_j; a value of $b_{ij} = 0$ represents that there is no precedence between the two parts c_i and c_j. If $i = j$, we set $b_{ij} = 0$. APM for a generator is shown as follows (Fig. 2):

	1	2	3	4	5	6	7	8	9	10	11	12	13	14	15
1	0	0	0	0	0	0	0	0	0	0	0	0	0	0	0
2	1	0	0	0	0	0	0	0	0	0	0	0	0	0	0
3	1	1	0	0	0	0	0	0	0	0	0	0	0	0	0
4	1	1	1	0	0	0	0	0	0	0	0	0	0	0	0
5	1	1	1	1	0	0	0	0	0	0	0	0	0	0	0
6	0	0	0	0	0	0	0	1	0	0	0	0	0	1	1
7	1	1	1	1	1	1	0	1	1	1	1	1	1	1	1
8	0	0	0	0	0	0	0	0	0	0	0	0	0	0	0
9	0	0	0	0	0	0	0	0	0	1	1	1	1	0	0
10	0	0	0	0	0	0	0	0	0	0	1	1	1	0	0
11	0	0	0	0	0	0	0	0	0	0	0	1	1	0	0
12	0	0	0	0	0	0	0	0	0	0	0	0	1	0	0
13	0	0	0	0	0	0	0	0	0	0	0	0	0	0	0
14	0	0	0	0	0	0	0	0	0	0	0	0	0	0	1
15	0	0	0	0	0	0	0	0	0	0	0	0	0	0	0

Fig. 2. Assembly precedence matrix for a generator

3.3   Employed Bee Phase

In the basic ABC algorithm, for each nectar source, there is only one employed bee. Employed bees use a greedy selection through its observed information to search for a better nectar source in the neighborhood of the present nectar source. To enrich the neighborhood structure and diversify of the population, simulated annealing algorithm are utilized to generate neighboring nectar sources for the employed bees [17].

The above method for generating new nectar sources follows the assembly precedence relations, thus it can improve the quality of nectar sources continually in the evolution process. ABC algorithm has strong exploration ability and weak exploitation ability. Therefore, a local search algorithm is presented in Sect. 3.6 in order to balance the global exploration and local exploitation of HABC. The selection of nectar sources carries out a greedy selection which is the same as in the basic ABC algorithm: if new nectar source is superior to its current nectar source in terms of profitability, the employed bee memorizes the new position and forgets the old one; otherwise, the previous position will be kept in memory.

3.4   Onlooker Bee Phase

Onlooker bees wait in the hive and select a nectar source to exploit with a certain probability based on the information shared by the employed bees. This selection is called the wheel selection which is easy to fall into local optimum. It maybe causes all the onlooker bees fly to the same nectar source, while the others cannot recruit onlooker bees. Instead of the wheel selection, we use the tournament selection due to its simplicity and ability to escape from local optimum.

In the tournament selection, an onlooker bee selects a nectar source in such a way that two nectar sources are picked randomly from the population and the better one is chosen. It means that half of the nectar sources will be exploited by onlooker bees and there is a greater probability to find the optimal nectar sources (the optimal or near-optimal assembly sequences). In addition, the onlooker utilizes the greedy selection as used by the employed bee. Then all the local nectar sources approach the local optimum, and this could help find better nectar sources. There are three options for the employed bee which has not recruited onlooker bees:

(a) It can return to the nectar source and continue to exploiting it.
(b) It can be an onlooker, waiting to be recruited by other employed bees in the hive.
(c) It can be a scout and starts searching around the hive for a new nectar source.

To guarantee the diversity of nectar sources and keep the nectar sources quantity constant at each iteration of the algorithm, we just consider the first and the third behaviors for such a bee.

3.5 Scout Bee Phase

In the basic ABC algorithm, if a nectar source cannot be further improved through a predetermined number of trials limit, then the nectar source is assumed to be abandoned, and the corresponding employed bee becomes a scout bee. To make full use of the swarm coordination of information sharing, we suppose that the scout bees learn from the current optimal (minimum fitness value) nectar sources when generating a new one. Since the search space around the best nectar source could be the most promising region, the new nectar source approaches the local optimum after every iteration of algorithm. This will increase the search efficiency and accelerate the converging speed in later convergence phase.

To avoid the algorithm trap into a local optimum, the scout bee first generates a nectar source randomly; then it learns from the current optimal nectar source with the probability P; finally, it can get a new nectar source. Next, the scout performs the function of the employed bee. The process of a scout generating a nectar source is that the new nectar source inherit the parts of the optimal nectar source with a random probability $r < P$, the rest inherit random sequences in order (Fig. 3).

3.6 A Local Search Algorithm

In order to enhance the exploitation capability of the algorithm, a simple local search is embedded in the proposed HABC algorithm. The local search is based on the insert and swap operator.

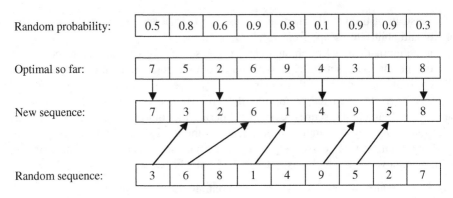

Fig. 3. The process of scout bee generate nectar source $(P = 0.7)$

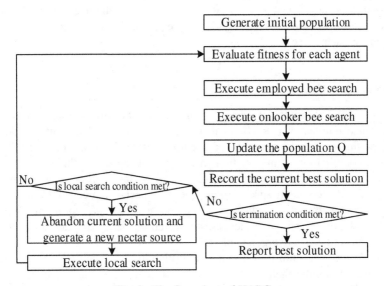

Fig. 4. The flow chart of HABC

3.7   Computational Procedure of the HABC Algorithm

The complete computational procedure is described in Fig. 4. Some parameters are need to set first: population size SN; learning rate P; local search probability P_L; the threshold value l_{max} of local search; round T_{max} of iteration. Then initial the population $Q = \{AS_1, AS_2, \ldots, AS_{SN}\}$, and half of Q are generated with the constraint APM and the rest are generated randomly. The employed bee search, onlooker bee search and scout bee phase which is also called local search and depend on the local search condition l_{max} are then executed, and Q is updated by previous steps and the current best solution is recorded. The algorithm is stopped based on the termination condition, that is, the round T_{max} of iteration. Finally, report the best solution.

4   Experimental Results and Analysis

The HABC algorithm for assembly sequence planning and optimization has been programmed with Visual C++ 6.0 language on a PC with 64-bit system of Windows 7 and 8 G memory. By comparing the HABC algorithm with the CPSO algorithm, firefly algorithm and improved firefly algorithm [18], the validity of this method is confirmed with actual assembly example.

4.1   Simulation Example

The exploded solid model of a generator assembly is shown in Fig. 5. The generator comprises 15 parts which are marked with numbers orderly. The assembly directions of all the parts are along the X and Y axes. Part 7 is viewed as the base part which has the most assembly relations with the other parts. Suppose that the assembly tool and assembly connector to assemble the base part 7 are T1 and C1 respectively. The assembly interference matrix, local preference matrix, stability matrix, connection matrix and tool matrix are given based on the assembly structure and assembly process. In this case, the proper tools and the connections for assembling each part are illustrated in Table 1, the assembly interference matrix are presented in Sect. 3.2 and the other assembly process constraints are omitted.

Table 1. Assembly tools and the connections for the parts

Part number	Assembly tool	Connection type
1	T_1, T_2, T_4	C_2
2	T_1, T_4, T_5	C_1
3	T_2, T_3, T_4	C_3
4	T_1, T_2, T_5	C_1
5	T_1, T_2, T_4	C_1
6	T_1, T_2, T_3	C_1
7	T_1	C_1
8	T_3, T_4, T_5	C_3
9	T_1, T_2, T_4	C_1
10	T_1, T_2, T_5	C_1
11	T_2, T_3, T_4	C_3
12	T_1, T_4, T_5	C_1
13	T_1, T_2, T_4	C_2
14	T_3, T_4, T_6	C_3
15	T_4, T_5, T_6	C_3

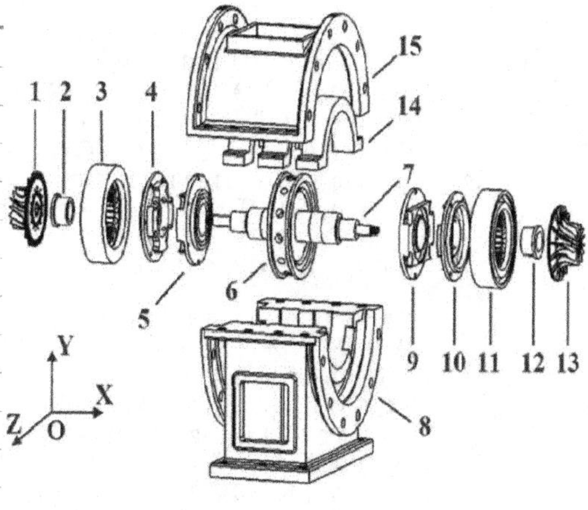

Fig. 5. The assembly of a generator

4.2   The Comparison Between Algorithms

By comparing the HABC algorithm with the CPSO algorithm, firefly algorithm and improved firefly algorithm, the validity of this method is confirmed. Under the same constraints, the results are shown in Tables 2 and 3 with the same indices and weights respectively. The other parameters of CPSO algorithm are assigned as literature [7], and the other parameters of firefly algorithm and improved firefly algorithm are as literature [6]. Since CPSO algorithm can obtain sub-optimal sequences only in larger cycles, we present results with 500 cycles. All the parameters of the assembly cost function and the HPSO algorithm are assigned as: $c_f = 50$, $c_s = 0.15$, $c_p = 0.15$, $c_d = 0.40$, $c_t = 0.15$, $c_c = 0.15$, $P = 0.7$, $P_L = 0.8$, $l_{max} = 5$.

Table 2. Comparison of the generated sequences of the four algorithms

Results		Sequence	n_f	n_s	n_p	n_d	n_t	n_c	f
HABC	1	7 6 5 4 9 10 11 12 13 3 2 1 14 15 8 +X +X +X +X −X −X −X −X −X +X +X +X +X +X +Y T_1 T_1 T_1 T_1 T_1 T_1 T_4 T_4 T_4 T_4 T_4 T_4 T_4 T_4 T_4 C_1 C_1 C_1 C_1 C_1 C_1 C_3 C_1 C_2 C_3 C_1 C_2 C_3 C_3 C_3	0	0	0	3	1	7	2.40
	2	7 6 9 10 5 4 3 2 1 11 12 13 14 15 8 −X −X −X −X +X +X +X +X +X −X −X −X −X −X +Y T_1 T_1 T_1 T_1 T_1 T_1 T_1 T_4 T_4 T_4 T_4 T_4 T_4 T_4 T_4 C_1 C_1 C_1 C_1 C_1 C_1 C_3 C_1 C_2 C_3 C_1 C_2 C_3 C_3 C_3	0	0	0	3	1	7	2.40
CPSO		7 6 5 4 9 10 11 12 3 2 1 13 14 15 8 +X +X +X +X −X −X −X −X +X +X +X −X −X −X +Y T_1 T_1 T_1 T_1 T_1 T_1 T_1 T_4 T_4 T_4 T_4 T_4 T_4 T_4 T_4 C_1 C_1 C_1 C_1 C_1 C_1 C_3 C_3 C_1 C_3 C_2 C_2 C_3 C_3 C_3	0	0	0	4	1	5	2.50
Firely algorithm	1	7 6 5 4 9 10 11 12 13 3 2 1 14 15 8 +X +X +X +X −X −X −X −X −X +X +X +X +X +X +Y T_1 T_1 T_1 T_1 T_1 T_1 T_1 T_4 T_4 T_4 T_4 T_4 T_4 T_4 T_4 C_1 C_1 C_1 C_1 C_1 C_1 C_3 C_1 C_2 C_3 C_1 C_2 C_3 C_3 C_3	0	0	0	3	1	7	2.40
	2	7 6 9 10 5 4 3 2 1 11 12 13 14 15 8 −X −X −X −X +X +X +X +X +X −X −X −X −X −X +Y T_1 T_1 T_1 T_1 T_1 T_1 T_4 T_4 T_4 T_4 T_4 T_4 T_4 T_4 T_4 C_1 C_1 C_1 C_1 C_1 C_1 C_3 C_1 C_2 C_3 C_1 C_2 C_3 C_3 C_3	0	0	0	3	1	7	2.40
Improved Firely algorithm	1	7 6 5 4 9 10 11 12 13 3 2 1 14 15 8 +X +X +X +X −X −X −X −X −X +X +X +X +X +X +Y T_1 T_1 T_1 T_1 T_1 T_1 T_1 T_4 T_4 T_4 T_4 T_4 T_4 T_4 T_4 C_1 C_1 C_1 C_1 C_1 C_1 C_3 C_1 C_2 C_3 C_1 C_2 C_3 C_3 C_3	0	0	0	3	1	7	2.40
	2	7 6 9 10 5 4 3 2 1 11 12 13 14 15 8 −X −X −X −X +X +X +X +X +X −X −X −X −X −X +Y T_1 T_1 T_1 T_1 T_1 T_1 T_1 T_4 T_4 T_4 T_4 T_4 T_4 T_4 T_4 C_1 C_1 C_1 C_1 C_1 C_1 C_3 C_1 C_2 C_3 C_1 C_2 C_3 C_3 C_3	0	0	0	3	1	7	2.40

As shown in Table 3, the HABC algorithm, firefly algorithm and improved firefly algorithm can all find two sequences with assembly cost of 2.40. The CPSO algorithm can only find the near-optimal assembly sequences with assembly cost of 2.50. HABC

Table 3. Comparison on running time (s)

Algorithm	Population	Cycles	c	Running time (s)
HABC	50	500	2.40	1.482
CPSO	50	500	2.50	4.652
Firefly algorithm	50	500	2.40	2.65
Improved firefly algorithm	50	500	2.40	65.29

algorithm has certain advantages compared with CPSO algorithm and firefly algorithm and the advantages are obvious compared with the improved firefly algorithm in the running time. Although the improved firefly algorithm has made improvements on firefly algorithm, the results of experiment in literature [15] showed that improved firefly algorithm is less efficient than firefly algorithm. This is also confirmed in this experiment.

Literature [6] concentrates on the convergence of firefly algorithm and improved firefly algorithm. Convergence reflects the approximation of the resulting solutions and the optimal solution at each iteration. In order to study the convergence of the HABC algorithm, we used the optimal solutions (the assembly cost of the example is 2.40) that the algorithm has found at the same iterations. In the same environment, the running results of the program are shown in Table 4. From the experimental results, convergence of the HABC algorithm has been relatively close to improved firefly algorithm and HABC algorithm only spent. A few one-tenth of the time of improved firefly algorithm.

Table 4. Comparison on convergence. (n_s is the number of optimal solutions)

Population Cycles	Algorithm	30		50	
		n_s	time(s)	n_s	time(s)
200	HABC	58	0.40	97	0.65
	Improved firefly algorithm	66	9.40	118	26.14
400	HABC	123	0.75	194	1.19
	Improved firefly algorithm	147	18.87	224	52.38

To sum up, the HABC algorithm that we proposed not only has outstanding search capability (in finding the optimal sequence) and efficiency (in time), but also improves the convergence rate while solving the assembly sequence planning problems. Therefore, we have confirmed that HABC algorithm can solve the assembly sequence planning very well.

5   Conclusions and Future Work

The HABC algorithm presented in this paper inherits the advantages of both the ABC algorithm and the simulated annealing algorithm. We encode the assembly sequences and discrete the nectar sources to make HABC algorithm suitable for the assembly sequence planning problems. Then, based on the strong exploration ability and weak exploitation ability of ABC algorithm, we use a local searching algorithm and a simulated annealing algorithm to enhance the ability of local exploitation. In the employed bee phase, we use the tournament selection instead of the wheel selection to escape from local optimum. In the scout bee phase, we design the scout bees to learn from the current optimal (minimum fitness value) nectar sources rather than randomly generated and this can overcome the defect of the slow convergence speed. According to the experiment, it is shown that the HABC algorithm is powerful for solving the assembly sequence planning problem.

In the HABC algorithm, there are some subjective operations in setting the weights of indexes that are closely related to the optimal solution. If some expert knowledge can be introduced and the interaction process can be analyzed by some professional method before setting the weights, then the results of ASP may have more guiding significance in practice. Therefore, the method of setting weights is one of our future works. Another future work is to extend the HABC algorithm from the assembly in orthogonal coordinate direction to the assembly in spherical coordinate.

Acknowledgments. This work is supported by the Natural Science Foundation of China (Nos. 61262030, 61572146, 61363030, U1501252); the Natural Science Foundation of Guangxi Province (Nos. 2015GXNSFAA139285, 2014GXNSFAA118354); the Guangxi Key Laboratory of Trusted Software, and the High Level of Innovation Team of Colleges and Universities in Guangxi and Outstanding Scholars Program.

References

1. Homem, M.A.: Correct and complete algorithm for the generation of mechanical assembly sequences. IEEE Trans. Robot. Autom. **2**(7), 228–240 (1991)
2. Gottipol, U.R.B., Ghos, K.: A simplified and efficient representation for evaluation and selection of assembly sequences. Comput. Ind. **50**(3), 251–264 (2003)
3. Wang, Y., Liu, J.H., Li, L.S.: Assembly sequence merging based on assembly unit partitioning. Int. J. Adv. Manuf. Technol. **45**, 808–820 (2009)
4. Greg, C.S., Shana, S.F.: An enhanced genetic algorithm for automated assembly planning. Robot. Comput. Integr. Manuf. **18**(2002), 355–364 (2000)
5. Chen, R.S., Lu, K.Y., Yu, S.C.: A hybrid genetic algorithm approach on multi-objective of assembly planning problem. Eng. Appl. Artif. Intell. **15**(4), 447–457 (2012)
6. Zeng, B., Li, M.F., Zhang, Y.: Assembly sequence planning based on improved firely algorithm. Comput. Integr. Manuf. Syst. **4**(20), 799–806 (2014). (in Chinese)
7. Wang, Y., Liu, J.H.: Chaotic particle swarm optimization for assembly sequence planning. Robot. Comput. Integr. Manuf. **26**(2010), 212–222 (2010)

8. Ibrahim, I., Ibrahim, Z., Ahmad, H., et al.: An assembly sequence planning approach with binary gravitational search algorithm. In: SoMeT, pp. 179–193 (2014)

9. Ibrahim, I., Ibrahim, Z., Ahmad, H., Yusof, Z.M.: An assembly sequence planning approach with a multi-state particle swarm optimization. In: Fujita, H., Ali, M., Selamat, A., Sasaki, J., Kurematsu, M. (eds.) IEA/AIE 2016. LNCS, vol. 9799, pp. 841–852. Springer, Heidelberg (2016). doi:10.1007/978-3-319-42007-3_71

10. Gao, W.F., Liu, S.Y., Huang, L.L.: Inspired artificial bee colony algorithm for global optimization problems. Acta Electronica Sinica **12**(40), 2396–2403 (2012)

11. Alatas, B.: Chaotic bee colony algorithms for global numerical optimization. Expert Syst. Appl. **37**(8), 5682–5687 (2010)

12. Bahriye, A., Karaboga, D.: A modified artificial bee colony algorithm for real-parameter optimization. Inf. Sci. **192**(1), 120–142 (2012)

13. Pan, Q.K., Suganthan, P.N., Chua, T.J.: A discrete artificial bee colony algorithm for the lot-steaming flow shop scheduling problem. Inf. Sci. **181**(2011), 2455–2468 (2011)

14. Percoco, G., Diella, M.: Preliminary evaluation of artificial bee colony algorithm when applied to multi-objective partial disassembly planning. Res. J. Appl. Sci. Eng. Technol. **6**(17), 3234–3243 (2013)

15. Lee, S.: Subassembly identification and evolution for assembly planning. IEEE Trans. Syst. Man Cybern. **24**(3), 493–503 (1994)

16. Tseng, Y.J., Chen, J.Y., Huang, F.Y.: A particle swarm optimization algorithm for multi-plant assembly sequence planning with integrated assembly sequence planning and plant assignment. Int. J. Prod. Res. **48**(10), 2765–2791 (2013)

17. Yuan, B.A., Zhang, C.Y., Shao, X.Y.: An effective hybrid honey bee mating optimization algorithm for balancing mixed-modal two-sided assembly lines. Comput. Oper. Res. **53** (2015), 32–41 (2015)

18. Zeng, B., Li, M.F., Zhang, Y.: Research on assembly sequence planning based on firely algorithm. J. Mech. Eng. **49**(11), 177–185 (2013). (in Chinese)

A Novel Track Initiation Method Based on Prior Motion Information and Hough Transform

Jun Liu[1], Yu Liu[1,2(✉)], and Wei Xiong[1]

[1] Research Institute of Information Fusion, Naval Aeronautical and Astronautical University,
Yantai 264001, China
18615042187@163.com, liuyu77360132@126.com, xiongwei@csif.org.cn
[2] School of Electronic and Information Engineering, Beihang University,
Beijing 100191, China

Abstract. This paper deals with the problem of track initiation in dense clutters. A novel Hough transform method based on prior motion information is proposed. Firstly, with time sequence of the measurements and targets' kinematic information considered, the single scan data accumulation effect is avoided, and the initiated tracks are more reliable. Then, measurements in observation space only vote to part of the lines that pass through them, which can effectively reduce the computational cost, and improve the initiation speed. Simulation results indicate that the proposed method is superior to the traditional algorithms in terms of detection probability, computational cost and robustness.

Keywords: Track initiation · Prior information · Hough transform · Target tracking

1 Introduction

Track initiation is the primary problem of multi-target tracking, especially for establishing the records for new targets and terminating the records of inexistent targets [1, 2]. Because of the long distance between the radar and targets, it is hard for the targets to be accurately measured by sensors, and there is no statistical rule to judge whether the measure is from target or clutter, track initiation itself is a very difficult issue. It is hoped that target tracks are initiated as early as possible, because the operators can get an early situation picture for the surveillance area. On the other hand, it is hoped that false track initiation rate is as low as possible. False tracks cause waste of radar resources, because radar beams for track maintenance are assigned to the area where there is no target [3]. In addition, to detect modern dim targets, radar needs to lower the threshold. Therefore, the false alarm rate will be very high, and how to deal with the track initiation of targets moving in such circumstance has been a challenging task now.

© IFIP International Federation for Information Processing 2016
Published by Springer International Publishing AG 2016. All Rights Reserved
Z. Shi et al. (Eds.): IIP 2016, IFIP AICT 486, pp. 72–77, 2016.
DOI: 10.1007/978-3-319-48390-0_8

2   Review of Current Techniques

During the past several decades, a lot of research has been contributed to track initiation, and there are two main series, one is sequential data processing technique, such as the heuristic rule method [4], logic-based method [1]; another is batch data processing technique, such as multiple hypothesis tracking method [5], integrated probabilistic data association algorithm [6], the Hough transform technique [7], and the modified Hough transform technique [8]. The former method has low computational cost, and is easy to apply. However, it is just suitable for initiation in sparse clutter. Though the latter method has a better performance than the former method, and is effective in reducing false alarm probability, it has much more computational cost than the former method, in addition, it is difficult to apply.

As is known, Hough Transform has low sensitivity to local fault and good ability in suppressing noise and clutters. Because of this advantage, it has been widely applied to track initiation in dense clutter environment. As a batch data processing method, Hough transform has the disadvantage of huge computation and memory. A lot of research has been done to reduce computation and memory by researchers, and many advanced methods have been put forward, such as modified Hough transform, random Hough transform, etc. A novel method for fast track initiation in dense clutters by divide and Hough transform method is proposed in this paper and this method is very effective in decreasing computation and memory. The measurement's time sequence information and targets' kinematics information also have been effectively utilized.

3   Our Approach

For Hough Transform, any point in data space is corresponding to the only one curve in parameter space, and points of the same line in data space are corresponding to different curves in parameter space, and those curves intersect at one point in parameters. If A_i, A_{i+1}, A_{i+2} denotes as an echo in the ith, $(i+1)$th, $(i+2)$th scan respectively, then A_i and A_{i+1} are connected to form a line $\overline{A_i A_{i+1}}$, which is corresponding to a point C_1 in parameter space with Hough Transform. In the same way, A_{i+1} and A_{i+2} are connected to form a line $\overline{A_{i+1} A_{i+2}}$, which is corresponding to a point C_2. in parameter space with Hough Transform. If A_i, A_{i+1}, A_{i+2} are collinear, C_1 and C_2 must coincide as shown in Fig. 1. If A_i, A_{i+1}, A_{i+2} are approximately collinear, then C_1 and C_2 will be very close, conversely, they will be far apart. According to this principle, a track can be detected through the intensive degree of point distribution in parameter space.

In this new method, any two observed data from different cycles is connected, which avoids loss of echo data. If only observed data from two adjacent cycles can be connected, when the target echo appears flashing, for example, a target echo is lost every other cycle, then echo data from two adjacent cycles can not be connected, which will cause loss of target information. However, not all observed data from two different cycles are from the same target, only when distance between the two points satisfies

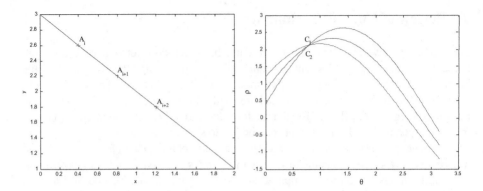

Fig. 1. Three collinear points and the corresponding curves

some condition, they can be from the same target. If $z_1(x_1, y_1)$, $z_2(x_2, y_2)$ is the measurement in time t_1, t_2 respectively, then distance between the two points is

$$d = \sqrt{(x_2 - x_1)^2 + (y_2 - y_1)^2} \tag{1}$$

where d is the distance. Then the average speed is

$$v = \frac{d}{|t_2 - t_1|} \tag{2}$$

where v denotes as the average speed. If v satisfies

$$v_{min} \le v \le v_{max} \tag{3}$$

z_1 and z_2 can be from the same target. In formula (3), v_{min} and v_{max} denote as the minimum and maximum speed of the target respectively. Through this process, the number of combinations is greatly reduced. If z_1 and z_2 satisfy the above condition, then the corresponding curve through Hough Transform in parameter space is

$$\begin{cases} \rho = x_1 \cos(\theta) + y_1 \sin(\theta), \\ \rho = x_2 \cos(\theta) + y_2 \sin(\theta). \end{cases} \tag{4}$$

By solving Eq. (4), the intersection of the line consisting of z_1 and z_2 in parameter space is

$$\tan \theta = -\frac{x_2 - x_1}{y_2 - y_1} \tag{5}$$

Let $a = -\dfrac{x_2 - x_1}{y_2 - y_1}$, because $\theta \in [0, \pi]$, then

$$\theta = \begin{cases} -\arctan(a) & a > 0, \\ -\arctan(a) + \pi & a \le 0. \end{cases} \tag{6}$$

Bring formula (6) back into Eq. (4), ρ is obtained.

$$\rho = \frac{|x_2 y_1 - x_1 y_2|}{d} \tag{7}$$

During the initiation period, targets are far away from sensors, so they are often thought to move in a straight line. According to the above discussion, any two measurements from a target must be mapped to the same (ρ, θ) in the parameter space. However, because of target's motion and measurement error, it is impossible that all (ρ, θ) is the same, but they will be concentrated in a small area. These small areas can be detected through the density of point distribution in the parameter space, and then the corresponding tracks are detected.

Suppose there are N intersections in parameter space, to detect target trajectory from observed data, the $\rho - \theta$ parameter space is equidistantly quantized into $N_\rho \times N_\theta$ cells with distance interval $\Delta\rho$ and angle interval $\Delta\theta$. Define parameter space array according to $\Delta\rho$ and $\Delta\theta$, if an intersection votes to accumulated cell $A(k, l), A(k, l) = A(k, l) + 1$. Search for the peak values in accumulated cells, and N_ξ denotes as the threshold. If $A(k, l)$ satisfies

$$A(k, l) > N_\xi \tag{8}$$

output the corresponding observed data as the initial tracks of targets.

4   Simulation and Results

In this section, suppose the surveillance area is 100000 m long and 100000 m wide. And assume 5 targets move in a straight line with constant velocity, and these targets are tracked with 2D radar. The initial position of 5 targets are (55000 m, 55000 m), (45000 m, 45000 m), (35000 m, 35000 m), (45000 m, 25000 m), (55000 m, 15000 m), and 5 targets move in the same speed, $v_x = 500$ m/s, $v_y = 0$ m/s. And assume radar's PRI is T = 5 s, azimuth observation square root error is $\sigma_\theta = 0.3°$, while distance observation square error is $\sigma_r = 40$ m. For a fair comparison, we make $N' = 100$ independent Monte Carlo runs, and the number of false measurements in the surveillance area satisfies the passion distribution with density λ.

Define the false track initiation probability as

$$P_F \triangleq \sum_{i=1}^{N'} f_i / \sum_{i=1}^{N'} n_i \tag{9}$$

where N' is the number of Monte Carlo simulation, f_i is number of false tracks in i^{th} Monte Carlo simulation, and n_i is number of all tracks in i^{th} Monte Carlo simulation.

In the same way, the detection probability of target t is given as

$$P_D = \frac{\sum\limits_{t=1}^{N_t} \sum\limits_{i=1}^{N'} l_{it}}{N_t \times N'} \tag{10}$$

where l_{it} indicates whether track of target t is correctly initiated in i^{th} Monte Carlo simulation, and l_{it} is defined as

$$l_{it} = \begin{cases} 1 & \text{if target } t \text{ is correctly initiated,} \\ 0 & \text{else.} \end{cases} \tag{11}$$

As is seen from Table 1, with the increase of clutters, when $\lambda = 100$, there are many false tracks in the former 4 algorithms. Especially in HT method, a great number of false tracks make it impossible to detect the exact tracks for targets. All these false tracks will cause a lot of trouble for subsequent target tracking and waste of radar resources. However, the method proposed in this paper is very robust to clutter interference, which can correctly initiate targets' tracks and suppress false tracks generated by clutters.

Table 1. Performance comparison of 5 track initiation methods

Track initiation method		P_D	P_F
$\lambda = 50$	The heuristic rule method	82.3 %	57.6 %
	The logic-based method	82.5 %	58.7 %
	The HT method	12.1 %	88.6 %
	The modified HT method	85.2 %	44.8 %
	The proposed method	89.3 %	38.5 %
$\lambda = 100$	The heuristic rule method	20.4 %	82.7 %
	The logic-based method	25.2 %	76.3 %
	The HT method	0.1 %	95.2 %
	The modified HT method	45.3 %	64.2 %
	The proposed method	63.2 %	42.1 %

For all these methods, the proposed method always keeps the highest probability of detection in both simulation scenarios. Even when $\lambda = 100$, it still keeps 63.2 %, while the highest detection probability of the remaining methods is 45.3 %. These simulation results indicate that the new method has a superior performance for track initiation, and is very robust to clutter interference.

To further validate the effectiveness of the proposed algorithm, describes the computational cost for methods based on Hough transform. From the table we know that the proposed method is less computational than the HT method and the modified HT method, and is more suitable for fast track initiation in dense clutters (Table 2).

Table 2. Comparison of computational cost for HT based methods

Track initiation method		Computational cost
$\lambda = 50$	The HT method	4.7×10^7
	The modified HT method	3.6×10^6
	The proposed method	2.1×10^4
$\lambda = 100$	The HT method	3.3×10^8
	The modified HT method	5.2×10^7
	The proposed method	8.6×10^5

5   Conclusions

In this paper, a novel method for Fast track initiation in dense clutters by divide and Hough transform method is proposed, which connects any two points from different cycles and calculates the exact intersection of the two points in parameter space. With the measurement's time sequence information and targets' kinematics information considered, this new algorithm avoids a single scan data accumulation effect, and can improve the reliability of formed tracks. In this new algorithm, measurements in observation space only vote to part lines instead of all in standard Hough transform that pass through them, so it can effectively reduce computation and memory, and improve the speed of track initiation. In addition, the simulation results also validate the superior performance of the new algorithm and indicate that it can effectively eliminate the false tracks, and is more robust to clutter inference.

Acknowledgements. This work is supported by the State Key Program for Basic Research of China (No. 613XXXXX), National Natural Science Foundation of China (No. 91538201).

References

1. Bar-Shalom, Y., Willett, P.K., Tian, X.: Tracking and Data Fusion: A Handbook of Algorithms. Academic, New York (2011)
2. Kim, S.-W., Lim, Y.-T., Song, T.-L.: A study of a new data association and track initiation method with normalized distance squared ordering. Int. J. Control Autom. Syst. **9**(5), 815–822 (2011)
3. Obata, Y., Maekawa, R., et al.: Track initiation algorithm for dim target using backward prediction. In: SICE Conference, pp. 1267–1273, Japan (2013)
4. Farina, A., Studer, F.A.: Radar Data Processing: Introduction and Tracking, vol. 1. Research Studies Press Ltd. (1985)
5. Blackman, S.S: Multiple hypothes for multiple target tracking. Aerosp. Electron. Syst. Mag. **19**(1), 5–18 (2004)
6. Musicki, D., Evans, R., Stankovic, S.: Integrated probabilistic data association. Automatic Control **39**(6), 1237–1241 (1994)
7. Casasent, D.P., Slaski, J: Optical track initiator for multitarget tracking. Appl. Opt. 27, 4546–4553 (1988)
8. Lo, T., et al.: Multitarget track initiation using a modified hough transform. Pointing Tracking Syst. **1**, 12–18 (1994)

Deep Learning

A Hybrid Architecture Based on CNN for Image Semantic Annotation

Yongzhe Zheng[1], Zhixin Li[1,2(✉)], and Canlong Zhang[1,2]

[1] Guangxi Key Lab of Multi-source Information Mining and Security,
Guangxi Normal University, Guilin 541004, China
zhengyzms@163.com, {lizx,clzhang}@gxnu.edu.cn
[2] Guangxi Experiment Center of Information Science, Guilin 541004, China

Abstract. Due to semantic gap, some image annotation models are not ideal in semantic learning. In order to bridge the gap between cross-modal data and improve the performance of image annotation, automatic image annotation has became an important research hotspots. In this paper, a hybrid approach is proposed to learn automatically semantic concepts of images, which is called Deep-CC. First we utilize the convolutional neural network for feature learning, instead of traditional methods of feature learning. Secondly, the ensembles of classifier chains (ECC) is trained based on obtained visual feature for semantic learning. The Deep-CC corresponds to generative model and discriminative model, respectively, which are trained individually. Deep-CC not only can learn better visual features, but also integrates correlations between labels, when it classifies images. The experimental results show that this approach performs for image semantic annotation more effectively and accurately.

Keywords: Semantic learning · Image auto-annotation · Convolutional neural network

1 Introduction

In the past decades, several state-of-the-art approaches have been proposed to solve the problems of automatic image annotation, which can be roughly categorized into two different models. The first one is based on generative model. The auto-annotation is first defined as a traditional supervised classification problem [1, 7], which mainly depends on similarity between visual features and predefined tags to model the classifier, then a unknown image is annotated relevant tags by computing similarity of visual level. The other is based on discriminative model, which are treat image and text as equivalent data. These methods try to mine the correlation between visual features and labels on an unsupervised basis by estimating the joint distribution of multi-instance features and words of each image [7, 16]. In brief, these methods extract various low-level visual features. These approaches greatly reduces the ability of feature presentation, therefore it makes the semantic gap become more serious between image and semantic.

Furthermore, the performances of image annotation are highly dependent on the representation of visual feature and semantic mapping. In view of the fact that deep

Z. Shi et al. (Eds.): IIP 2016, IFIP AICT 486, pp. 81–90, 2016.
DOI: 10.1007/978-3-319-48390-0_9

convolutional neural networks (CNNs) has been demonstrated a outstanding performance in computer vision recently. Besides, Mahendran and Vedaldi [3, 4, 9] and [11] have demonstrated that CNN has a better effect over existing methods of hand-crafted features in many applications, such as object classification, face recognition, and image annotation. Inspired these articles, this paper proposes a hybrid architecture based on CNN for image semantic annotation to improve the performances of image annotation.

In this paper, our main contributions are the following. Firstly, we use redesigned CNN model to learn high-level visual features. Secondly, we employ the ensembles of classifier chains (ECC) to train model on visual features and predefined tags. Finally, we propose a hybrid framework to learn semantic concepts of images based CNN (Deep-CC). Deep-CC not only can learn better visual features, but also integrates correlations between labels when it classifies images. The experimental results show that our approach performs more effectively and accurately.

2   CNN Visual Feature Extraction

In the past few years, some recent articles [14, 17] have demonstrated that the CNN models pre-trained on large datasets with data diversity, e.g., AlexNet [4] which can be directly transferred to extract CNN visual features for various visual recognition tasks such as image classification and object detection. CNN is a special form of neural network that consists of different types of layers, such as convolutional layers, spatial pooling layers, local response normalization layers and fully connected layers. Different network structures will show different ability of visual features representation. Krizhev et al. [4] have proved that the Rectified Linear Units (ReLUs) not only saves the computing time, but also implements the features of sparse representation, and ReLU also increases the sample characteristic diversity. So in order to improve the generalization ability of the feature representation, we extract fc7 visual vectors after ReLU. As shown in the top of the Fig. 1, our CNN model has the similar network structure to the AlexNet. As reflected in Fig. 1, which contains five convolutional layers (short as conv) and three fully-connected layers (short as fc). The CNN model is pre-trained in 1.2 million images of 1000 categories from ImageNet [14].

2.1   Extracting Visual Features from Pre-trained CNN Model

Li and Yu [5] and Razavian et al. [12] have demonstrated the outstanding performance of the off-the-shelf CNN visual features in various recognition tasks, so we utilize the pre-trained CNN model to extract visual features. Particularly, each image is resized to 227 * 227 and fed into the CNN model. As shown in Fig. 1, it represents the feature flow extracted from the convolutional neural network. The $fc7$ features are extracted from the secondly convolution layer after ReLU. The $fc7$ denote the 4096 dimensional features of the last two fully-connected layers after the rectified linear units (ReLU) [4].

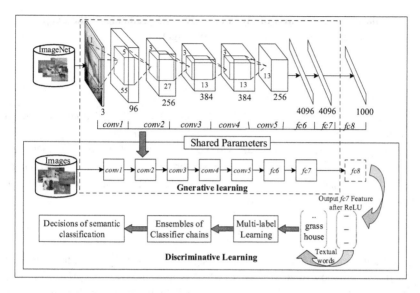

Fig. 1. The Pipeline of Image Annotation

2.2 Exacting Visual Feature from Fine-Tuned CNN Model

Taking into account the different categories (and the number of categories) between the target dataset and ImageNet, if we directly utilize the pre-trained model on the ImageNet to exact image visual features, it may not be the optimum strategy. To make the model fit the parameters better, we redesign the last hidden layer for visual feature learning task, later re-designed CNN model by fine-tuning parameters with each of images in the target dataset. Considering the rationality of the design of the convolutional neural networks, our CNN model has the similar network structure to the AlexNet. As show in the mid of Fig. 1, the overall architecture of our CNN model still contains five conv layers, followed by a pooling layer and three fully-connected layers. We redesign the last hidden layer for feature learning task. The number of neural units of the last fully-connected layer is modified from 1000 to m, where m is the number of the target dataset's categories. The output of the last fully-connected layer is then fed into a m-way softmax which produces a probability distribution over m categories.

Given one training sample x, the network extracts layer-wise representations from the first conv layer to the output of the last fully connected layer $fc8 \in \mathbb{R}^m$, which can be viewed as high level features of the input image. Followed by a softmax layer, $fc8$ is transformed into a probability distribution $p \in \mathbb{R}^m$ for objects of m categories, and cross entropy is used to measure the prediction loss of the network. Specifically, we define the following formula.

$$p_i = \frac{\exp(\hat{v}_i)}{\sum_i \exp(\hat{v}_i)} \text{ and } L = t_i \log(p_i) \tag{1}$$

In formula (1), L is the loss of cross entropy. The gradients of the deep convolutional neural network is calculated via back propagation

$$\frac{\partial L}{\partial \hat{h}_i} = p_i - t_i \tag{2}$$

In formula (2), $t = \{t_i | t_i \in \{0, 1\}, i = 1, \ldots, m, \sum_{k=1}^{m} t_i = 1\}$ denotes the true label of the sample x_j, where the $\{x_j | j = 1, 2, \ldots, n\}$ is a bag of instances. $t = \{t_i | t_i \in \{0, 1\}, i = 1, \ldots, m.\}$ is the label of the bag; Convolutional neural network extracts representations of the bag, it can get a feature vector $v = \{v_{ij}\} \in \mathbb{R}^{m \times n}$, in which each column is the representation of an instance. The aggregated representation of the bag for visual vectors are defined as follows.

$$\hat{v}_i = f(v_{i1}, v_{i2}, \ldots, v_{in}) \tag{3}$$

In the training phase, similarly back propagation algorithm is used to optimize the loss function L. Suppose that we have a set of training images $I = \{M_i\}$. The trained instances of traditional supervised learning in which training instances are given as pairs $\{(m_i, l_i)\}$, where $m_i \in \mathbb{R}^m$ is a feature vector and $l_i \in \{0, 1\}$ is the corresponding label. In visual feature learning, trained sample is regarded as bags $\{I_i\}$, and there are a number of instances x_{ij} in each bag. Finally, the network extracts layer-wise representations from the first conv layer to the output of the last fully connected layer visual vectors v_i, which can be viewed as high level features of the input image. By fine-tuning like this, the parameters can better adapt to the target dataset by rectifying the transferred parameters. For the task of visual feature learning, we first employ existing model to fine-tune the parameters in the target dataset, then we apply the fine-tuned CNN model to learn image visual features. Similarly, the FT-*fc7* denotes the 4096 dimensional features of the last two fully-connected layers after the rectified linear units (ReLU).

3 Ensembles of Classification Classifiers for Semantic Learning

In the discriminative learning phase, the ensemble of classification classifiers (ECC) [13] are used to accomplish the task of multi label classification, and each of the binary classifier is implemented by SVM. Taking into account the semantic correlations between tags, ECC can classify images into multiple semantic classes, with a high degree of confidence and acceptable computational complexity. Furthermore, by learning the semantic relevance between labels, classifier chain can effectively overcome the problems of label independence in image binary classification.

The classifier chain model consists of $|L|$ binary classifiers, where L denotes the truth label set. Classifiers are linked along a chain where each classifier deals with the binary relevance problem associated with label $l_j \in L$. The feature space of each linked in the chain is extended with the $\{0, 1\}$ label associations of all previous links. The training procedure is outlined in Algorithm 1 in the left of Table 1. Lastly, we can note the notation for a training example (x, S), where $S \subseteq L$ and x is an instance feature vector.

Hence a chain $C_1, C_2, \ldots, C_i$ of binary classifier is formed. Each classifier C_j in the chain is responsible for learning and predicting the binary association of label l_j, which is given in the feature space and is augmented by all prior binary relevance predictions in the chain $\{l_1, l_2, \ldots, l_{j-1}\}$. The classification procedure begins at C_1 and propagates along the chain C_1 determines $\Pr(l_1|x)$ and every following classifier $C_2, \ldots, C_j$ predicts $\Pr(l_j|x_i, l_1, l_2, \ldots, l_{j-1})$. This classification procedure is described in Algorithm 2 in the right of Table 1.

This training method passes label information between classifiers, with classifier chain taken into account label correlations, so it overcomes the label independence problem of binary relevance method. However, classifier chain still remains advantages of binary relevance method including low memory and runtime complexity. Although $|L|/2$ features are added to each instance on an average, this item is negligible in computational complexity because $|L|$ is invariably limited in practice.

Different order of the chain clearly has a different effect on accuracy. This problem can be solved by using an ensemble framework with a different random train ordering for each iteration. Ensembles of classifier chains train m classifier chains $C_1, C_2, \ldots, C_m$. Each C_k model is trained with a random chain which can order the L outputs and get a random subset of D. Hence each C_k model is likely to be unique and able to give different multi-label predictions. These predictions are summed by label so that each label receives a number of votes. A threshold is used to select the most popular labels which form the final predicted multi-label set. These predictions are summed by label so that each label receives a number of votes. A threshold is used to select the most popular labels which form the final predicted multi-label set.

Table 1. Training and prediction procedures of ensembles of classifier chains for multi-label learning

Processing	Algorithm 1. Training steps of classifier chain	Algorithm 2. Classifying procedure ECC
Input	Training set $I = \{(x_1, S_1), (x_2, S_2), \ldots, (x_n, S_n)\}$	Test example x.
Output	Classifier chains $\{C_1, C_2, \ldots, C_{|L|}\}$	$Y = \{l_1, l_2, \ldots, l_{|L|}\}$.
procedures		
1	For $i \in 1, 2, \ldots |L|$	$Y \leftarrow \{\}$
2	Semantic learning	For $i \in 1, 2, \ldots, |L|$
3	$I' \leftarrow \{\}$	Do $Y \leftarrow Y \cup (l_i \leftarrow C_i: (x_i, l_1, l_2, \ldots l_{j-1}))$
4	For $(x, S) \in I$	Return (x, Y)
5	Do $I' \leftarrow I' \cup ((x, l_1, l_2, \ldots, l_{i-1}), l_i)$	
6	Train C_i to predict binary relevance of l_i	
7	$C_i: I' \rightarrow l_i \in \{0,1\}$	

Each kth individual model predicts vector $y_k = (l_1, l_2, \ldots, l_{|L|}) \in \{0, 1\}^{|L|}$. The sums are stored in a vector $W = (\lambda_1, \lambda_2, \ldots, \lambda_{|L|}) \in \mathbb{R}^{|L|}$, where λ_j is defined as $\lambda_j = \sum_{k=1}^{m} l_j \in y_k$. Hence each $\lambda_j \in W$ represents the sum of the votes for label $l_j \in L$. We then normalize W to W_{norm}, which represents a distribution of scores for each label in $[0,1]$. A threshold is used to choose the final multi-label set Y such that $l_j \in Y$ where $\lambda_j \geq t$ for threshold t. Hence the relevant labels in Y represent the final multi-label prediction.

4   Hybrid Framework for Image Annotation

On the deep model and ensembles of classifier chains, we propose a hybrid learning framework to address cross-modal semantic annotation problem between images and text with Multi-label. Figure 2 shows two setups of the hybrid architecture approach for semantic learning based on deep learning. The first path (generative learning) feeds training image to the fine-tune pre-trained CNN step which is also called the feature learning phase, then in the discriminative learning phase, we utilize ensemble of classifier chains to model the visual vectors which are co-occurrence matrix consisting of texture and exacted visual features by pre-trained CNN model. This hybrid pipeline model is called Deep-CC image annotation system.

Bases on the learning feature, the trained CNN model output visual features after ReLU. Suppose that we have a set of images $M = \{m_1, m_2, \ldots, m_i\}$, this model extracts visual vectors by pre-trained CNN model and we denote the space of visual vectors as $V = \{v_1, v_2, \ldots, v_i\}$, where v_i denotes the visual vector of ith image. Noting the notation for a training example (v_i, S), where $S \in L$, L denotes the label set and v is a feature vector. Then, by making use of the aspect distribution and original labels of each training image, we build a series of classifiers in which every word in the vocabulary is treated as an independent class. The classifier chain model implements the feature classification task and it can effectively learn the semantic correlation between labels in discriminative step. Finally, given a test image, the Deep-CC system will return a correlative label subset $l \in L$.

Image					
Ground Truth	temple, sky, buddhist, mountains	elephant, trees, planes, sky	cabin, trees, autumn, field	trees, sky, road, park	people, water, trees, sand
HGDM annotations	house, sky, water, cloud, mountains	Africa, sky, land, animal, beach	field, grass, land, trees, mountains	trees, road, sky, pant, sea	sand, water, tress, people, cloud
Deep-CC annotations	temple, sky, palace, land, mountains	elephant, land, plane, trees, sky	chair, trees, field, land, mountains	park, road, trees, sky , mountains	sand, coast, trees, water, people

Fig. 2. Comparison of annotations made by HGDM and Deep-CC on Corel5k

As a comparison, we evaluate the deep feature's performance from the AlexNet CNN on those same benchmarks. Following by [10], we choose 5 words with highest confidence as annotations of the test image. After each image in the database is annotated, the retrieval algorithm ranks the images labeled with the query word by decreasing confidence.

5 Experiments and Results

In this section, we conduct experiments of our Deep-CC learning framework on both image classification and image auto-annotation. We choose a dataset Corel5K which is widely used in image classification and annotation. In order to make the experimental result more convinced, we simultaneously compare the experimental results with the existing traditional model and deep model.

5.1 Datasets and Evaluation Measures

In order to test the effectiveness and accuracy of the proposed approach, we conduct our experiments on a baseline annotated image datasets Corel5K [2]. Corel5k is a basic comparative dataset for recent research works on image annotation. The dataset contains 5000 images from 50 Corel Stock Photo cds. We divided this dataset into 3 parts: a training set of 4000 images, a validation set of 500 images and a test set of 500 images. Like the Duygulu et al. [2], We divide separately the training set of 4500 images and the test set of 500 images.

Image annotation performance is evaluated by comparing the captions automatically generated for the test set with the human-produced ground truth. It is essential to include several evaluation measures in multi-label evaluation. Similar to Monay and Gatica-Perez [10], we use mAP as evaluation measures. Naturally, we define the automatic annotation as the top 5 semantic words of largest posterior probability, and compute the recall and precision of every word in the test set.

5.2 Results for Image Annotation on Corel5 K

In this section, we demonstrate the performance of our model on the corel5 k data set for image multi-label annotation, and compare the results with some existing image annotation methods, e.g. PLSA-WORDS [10], HGMD [8] and DNN [15]. We evaluate the returned keywords in a class-wise manner. The performance of image annotation is evaluated by comparing the captions automatically generated with the original manual annotations. Similar to Monay and Gatica-Perez [10], we compute the recall and precision of every word in the test set and use the mean of these values to summarize the system performance.

Table 2 reports results of several models on the set of all 260 words which occur in the training set. Data in precision and recall columns denotes mean precision and mean recall of each word. The off-the-shelf CNN features (i.e. fc7 and FT-fc7) obtain significant improvements (7.8 % based on PLSA-WORDS, 3.4 % based on HGDM) compared with these traditional feature learning methods. After fine-tuning, a further improvement (8.2 % based on PLSA-WORDS, 4.6 % based on HGDM) can be achieved with the best performance of the CNN visual features FT-fc7.

Annotations of several images obtained by our Deep-CC annotation system are show in Fig. 2. We can see that annotations generated by Deep-CC are more accurate than HGDM in most cases. In order to be more intuitive to observe different precision-recall in various methods, the Fig. 3 presents the precision-recall curves of several annotation models on the Corel5k data set. As is shown in Fig. 3, Deep-CC performs consistently better than other models. Where the precision and recall values are the mean values calculated over all words.

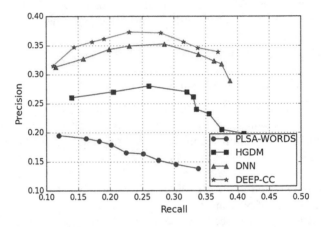

Fig. 3. Precision–recall curves of several models for image annotation on Corel5K

5.3   Result Analysis

In summary, the experimental results on Corel5k show that Deep-CC outperforms many state-of-the-art approaches, which proves that the redesigned CNN and the hybrid framework is effective in learning visual features and semantic concepts of images. We compare the CNN visual features with traditional visual features for learning semantic concepts of images over two traditional learning approaches and a deep model. Especially, the comparison in terms of rigid and articulated visual features among Corel5k is shown in Table 2, from which it can be seen that CNN feature outperforms almost all the original hand-crafted features. To verify this assumption, different visual features between traditional models (also from the authors of this paper) and CNN mode, and FT-fc7 is executed to make an enhanced prediction for Corel5k. Incredibly, the mAP score on Corel5k can surge to 35.2 % as shown in Table 2, which demonstrates the great dominance in the deep networks. To sum up, based on the above reported experimental results, we can see that CNN visual features are very effective for semantic image annotation.

Table 2. Performance (mAP in %) comparison in terms of different methods and visual features algorithms on Corel5k. (The "-" means to use their method)

Method	Visual features	Result on all words		mAP
		Precision	Recall	
PLSA-WORDs	-	22.1	12.1	19.1
	Fc7	27.5	21.7	26.9
	FT-fc7	29.3	22.6	27.3
HGDM	-	32.1	29.3	26.3
	Fc7	36.4	30.5	29.7
	FT-fc7	37.6	32.9	30.9
DNN	–	37.5	40.5	32.7
Deep-CC(our)	–	39.7	37.6	35.2

6   Conclusion

In this paper, we utilize CNN model to learn deep visual features, and we redesign the last hidden layer for feature learning task, and in order to obtain high performance of feature representation, we first train our deep model on ImageNet, then the pre-trained parameters are fine-tuned on target dataset. We showed under what conditions each visual feature can perform better, and propose a hybrid architecture. We demonstrated that re-designed CNN model and ensembles of classifier chains can effectively improve annotation accuracy.

In comparison to many state-of-the-art approaches, experimental results show that our method achieves superior results in the tasks of image classification and annotation on Corel5K. However, in the process of learning visual features, Deep-CC only employ single convolution neural network not fully understanding multiple instance in the image, and how to excavate the high-level semantic relevance between the tags, it can be deeply studied. In future research, we aim to take semi-supervised learning based on a large number of unlabeled data to improve its effectiveness.

Acknowledgement. This work is supported by the National Natural Science Foundation of China (Nos. 61663004, 61262005, 61363035), the Guangxi Natural Science Foundation (2013GXNSFAA019345, 2014GXNSFAA118368), the Guangxi "Bagui Scholar" Teams for Innovation and Research Project, Guangxi Collaborative Innovation Center of Multi-source Information Integration and Intelligent Processing.

References

1. Cusano, C., Ciocca, G., Schettini, R.: Image annotation using SVM. In: Proceedings of SPIE - The International Society for Optical Engineering, vol. 5304, pp. 330–338 (2003)
2. Duygulu, P., Barnard, K., de Freitas, J.F.G., Forsyth, D.: Object recognition as machine translation: learning a lexicon for a fixed image vocabulary. In: Heyden, A., Sparr, G., Nielsen, M., Johansen, P. (eds.) ECCV 2002, Part IV. LNCS, vol. 2353, pp. 97–112. Springer, Heidelberg (2002)

3. He, K., Zhang, X., Ren, S., et al.: Deep residual learning for image recognition. In: 2016 IEEE Conference on Computer Vision and Pattern Recognition (CVPR). IEEE (2016)

4. Krizhevsky, A., Sutskever, I., Hinton, G.E.: ImageNet classification with deep convolutional neural networks. In: Advances in Neural Information Processing Systems, vol. 25, no. 2 (2012)

5. Li, G., Yu, Y.: Visual saliency based on multiscale deep features. In: 2015 IEEE Conference on Computer Vision and Pattern Recognition (CVPR). IEEE (2015)

6. Li, J., Wang, J.Z.: Automatic linguistic indexing of pictures by a statistical modeling approach. IEEE Trans. Pattern Anal. Machine Intell. 25(9), 1075–1088 (2003)

7. Liu, Y., Zhang, D., Lu, G., et al.: A survey of content-based image retrieval with high-level semantics. Pattern Recogn. 40(1), 262–282 (2007)

8. Li, Z., Shi, Z., Zhao, W., et al.: Learning semantic concepts from image database with hybrid generative/discriminative approach. Eng. Appl. Artif. Intell. 26(9), 2143–2152 (2013)

9. Mahendran, A., Vedaldi, A.: Understanding deep image representations by inverting them. Comput. Sci., 5188–5196 (2015)

10. Monay, F., Gatica-Perez, D.: Modeling semantic aspects for cross-media image indexing. IEEE Trans. Pattern Anal. Mach. Intell. 29(10), 1802–1817 (2007)

11. Mnih, V., Kavukcuoglu, K., Silver, D., et al.: Human-level control through deep reinforcement learning. Nature 518(7540), 529–533 (2015)

12. Sharif Razavian, A., Azizpour, H., Sullivan, J., et al.: CNN features off-the-shelf: an astounding baseline for recognition. In: IEEE Conference on Computer Vision and Pattern Recognition Workshops, pp. 512–519. IEEE Computer Society (2014)

13. Read, J., Pfahringer, B., Holmes, G., et al.: Classifier chains for multi-label classification. Mach. Learn. 85(3), 254–269 (2011)

14. Russakovsky, O., Deng, J., Su, H., et al.: ImageNet large scale visual recognition challenge. Int. J. Comput. Vis. 115(3), 211–252 (2015)

15. Sermanet, P., Eigen, D., Zhang, X., et al.: OverFeat: integrated recognition, localization and detection using convolutional networks. Eprint Arxiv (2013)

16. Smeulders, A.W.M., Worring, M., Santini, S., et al.: Content-based image retrieval at the end of the early. IEEE Trans. Pattern Anal. Mach. Intell. 22(12), 1349–1380 (2000)

17. Wu, J., Yu, Y., Chang, H., et al.: Deep multiple instance learning for image classification and auto-annotation. In: Computer Vision and Pattern Recognition, pp. 3460–3469. IEEE (2015)

Convolutional Neural Networks Optimized by Logistic Regression Model

Bo Yang, Zuopeng Zhao, and Xinzheng Xu[(✉)]

School of Computer Science and Technology,
China University of Mining and Technology, Xuzhou 221116, China
xuxinzh@163.com

Abstract. In recent years, convolutional neural networks have been widely used, especially in the field of large scale image processing. This paper mainly introduces the application of two kinds of logistic regression classifier in the convolutional neural network. The first classifier is a logistic regression classifier, which is a classifier for two classification problems, but it can also be used for multi-classification problems. The second kind of classifier is a multi-classification logistic regression classifier, also known as softmax regression classifier. Two kinds of classifiers have achieved good results in MNIST handwritten digit recognition.

Keywords: Convolution neural network · Logistic regression · Softmax regression

1 Introduction

In recent years, since the convolutional neural network is proposed [1], it is widely used in pattern recognition [2], image processing [3], especially achieved good results in the large field of image processing [4]. This paper [5] makes a detailed theoretical analysis of the convolution neural network, then the various classification algorithms and models have been proposed. In the paper [6] proposed the multilayer perceptron as a convolutional neural network classifier, the paper [7] also used k nearest neighbor algorithm as a convolutional neural network classifier, the paper [8] used support vector machine (SVM) as the convolutional neural network classifier, both of them have achieved good results in the handwritten numeral recognition experiment. The paper [9, 10] mainly introduces the model of linear regression, logistic regression and so on. The paper [11] introduces the softmax regression model, and the detailed formula derivation of the algorithm combined with the back propagation algorithm.

The structure of this paper is divided into five parts. The second part mainly introduces the convolutional neural network structure. The third part mainly introduces two kinds of classification model. The first is a logistic regression model and how to use logistic regression to solve multi classification problem, on the other is the Soft regression model. The forth part of the thesis is the experiment and the result analysis. The last part of the paper is the summary.

© IFIP International Federation for Information Processing 2016
Published by Springer International Publishing AG 2016. All Rights Reserved
Z. Shi et al. (Eds.): IIP 2016, IFIP AICT 486, pp. 91–96, 2016.
DOI: 10.1007/978-3-319-48390-0_10

2   Structure of Convolutional Neural Network

Convolutional neural network can be functionally divided into two parts, one is the image feature extraction section, the other part of the classifier. In our experiment, convolutional neural network in the structure can be divided into seven layers including an input layer, convolutional layer C1, sub-sampling layer S2, convolutional layer C3, sub-sampling layer S4, sub-sampling layer S4 unfolded layer (not included in the number of layers), fully connected layer and output layer. The feature extraction part includes C1, S2, C3, S4 layer, the classifier part includes the sub-sampling layer S4 expansion layer (also can be used as the input layer of the classifier), fully connected layer and output layer.

Convolutional neural network structure has many variants, this article uses a classic convolution neural network. In this paper, we mainly introduce the application of two kinds of logistic regression classifiers in the convolutional neural network, so we simply introduce the convolution neural network. We can more detailed understanding of convolutional neural networks in this classic paper [5]. Figure 1 is the structure of the convolution neural network used in this paper. This paper carries out a handwritten numeral recognition experiment on the MNIST data set.

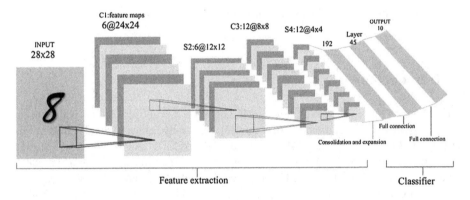

Fig. 1. Network structure of convolutional neural network

3   Classifier Model Based on Logistic Regression

Logistic regression is a kind of classifier for two classification problems. If it is used for multi classification problem, we need to train a logistic regression classifier for each category. The Softmax regression classifier is applied to solve the multi classification problem. The image feature extracted from the convolutional neural network is used as the input layer of the multi-layer neural network, and the classifier is used to classify the image, and the back propagation algorithm is used to update the weight parameters in the whole training process. Next, we will introduce these two kinds of classifiers in detail.

3.1 Logistic Regression Model

The two classification problem is divided into C_1 and C_2 two categories, then the output layer has only one neuron and its output is y. Assuming that the number of weight parameters of the output layer is n, then $w^L = [w_1, w_2, \ldots, w_n]$. Since the output value y is processed by the sigmoid function, and the output range of the sigmoid function is 0 to 1, then we can put the output value y. as the probability value which the input vector x^1 belongs to the first category C_1. Conversely, using one minus the output y gives the probability value that the input vector x^1 belongs to the second category C_2.

$$p\left(C_1|x^1\right) = y \tag{1}$$

$$p\left(C_2|x^1\right) = 1 - y \tag{2}$$

For the two classification problem with N training samples, the cost function E can be expressed as:

$$E = \frac{1}{N} \sum_{i=1}^{N} \left[-z^{(i)} \log\left(y^{(i)}\right) - (1 - z^{(i)})\log(1 - y^{(i)})\right] \tag{3}$$

If we use logistic regression to solve the multi classification problem, we need to train a logistic regression classifier for each category. For the K classification problem, then the output layer has K neural units, the use of logistic regression to solve the problem of K classification need to train K logical regression classifier. For a specific category, the rest of the category as a category, then it becomes a two classification problem. For the K classification problem with N training samples, the cost function E can be expressed as:

$$E = \frac{1}{N} \sum_{i=1}^{N} \sum_{k=1}^{K} \left[-Z_k^{(i)} \log\left(y_k^{(i)}\right) - (1 - Z_k^{(i)})\log(1 - y_k^{(i)})\right] \tag{4}$$

In order to prevent over fitting in the training process of neural network, we can add the regularization in the cost function. Regularization controls the model's complexity and makes the model more generalizable to unseen data. In order to simplify the problem, it is assumed that the neural network model is only the input layer and the output layer. We assume that the number of neural units in the input layer is M, while the output layer has K neural units. The weights between the input layer and the output layer are w, then the cost function is as follows:

$$E = \frac{1}{N} \sum_{i=1}^{N} \sum_{k=1}^{K} \left[-Z_k^{(i)} \log\left(y_k^{(i)}\right) - (1 - Z_k^{(i)})\log(1 - y_k^{(i)})\right] + \frac{\lambda}{2} \sum_{m=1}^{M} \sum_{k=1}^{K} w_{mk}^2 \tag{5}$$

In the above model, we only add the L2 regularization. The w_{mk} represents the connection weights of neural unit m and neural unit k.

3.2   Softmax Regression Model

We need to use the softmax function instead of the activation function in solving the K classification problem and obtain the class probabilities:

$$p\left(C_k|x^1\right) = y_k = \frac{e^{u_k}}{\sum_{i=1}^{K} e^{u_i}} \tag{6}$$

This is called as a multinomial logit model For the target vector z to meet the relationship $\sum_{i=1}^{K} z_i = 1$. Assuming K = 5, if the correct classification of a training sample is C_3, then the target vector should be $z^{(i)} = [0, 0, 1, 0, 0]^T$. Then the probability formula of the input vector x^1 and the target output vector z is:

$$p\left(z|x^1\right) = \prod_{k=1}^{K} y_k^{z_k} \tag{7}$$

For the multi classification problem with N training samples, the cost function of the Softmax regression model is:

$$E = -\frac{1}{N} \sum_{i=1}^{N} \sum_{k=1}^{K} z_k^{(i)} log y_k^{(i)} \tag{8}$$

Because the model has the redundant weight parameters, we need to add a weight decay term which penalizes large values of parameters to modify the cost function, our cost function is now:

$$E = -\frac{1}{N} \sum_{i=1}^{N} \sum_{k=1}^{K} z_k^{(i)} log y_k^{(i)} + \frac{\lambda}{2} \sum_{m=1}^{M} \sum_{k=1}^{K} w_{mk}^2 \tag{9}$$

3.3   Back Propagation

The "errors" which we propagate backwards through the network can be thought of as "sensitivities" of each unit with respect to perturbations of the bias [5]. That is to say,

$$\delta = \frac{\partial E}{\partial b} = \frac{\partial E}{\partial u} \frac{\partial u}{\partial b} \tag{10}$$

As a result of $\frac{\partial u}{\partial b} = 1$, the sensitivity of the output layer can be obtained.

$$\delta^L = \frac{\partial E}{\partial u^L} = y^{(i)} - z^{(i)} \tag{11}$$

By deriving the formula, we can obtain the sensitivity of the previous layer as follows:

$$\delta^l = (w^L)^T \delta^L \cdot f'(u^l) \tag{12}$$

Due to the use of the back propagation algorithm, we also need to know the gradient in the process of updating the weight parameters. The gradient formula is as follows:

$$\frac{\partial E}{\partial w^l} = \frac{\partial E}{\partial u^l}\frac{\partial u^l}{\partial w^l} = \delta^l(x^l)^T + \frac{\lambda}{N}w^l \tag{13}$$

$$\Delta w^l = -\eta\frac{\partial E}{\partial w^l} \tag{14}$$

4 Experiment and Result Analysis

We have made a good effect on the handwritten digit recognition [12] based on MNIST data set [13] with two kinds of logic regression classifier combined with convolutional neural network. We use three kinds of classifiers to do the contrast experiment, one is based on the logistic regression classifier, one is the softmax regression classifier, and the last one is the multilayer perceptron. We find that whether the classifier uses the hidden layer has a great influence on the convergence speed and the test results. The experimental results of the classifier without using the hidden layer are as follows (Table 1):

Table 1. Comparison of experimental results (error rate)

Method \ Iterations	CNN+MLP	CNN + LRM	CNN + Softmax
1	11.15%	14.96%	5.71%-7.49%
5	4.34%	8.01%	2.67%-3.26%
10	2.73%	5.89%	2.37% -2.75%
50	1.51%	3.24%	1.69%-2.68%
100	1.21%	2.27%	1.13%-1.61%

The experimental results of the classifier using the hidden layer are as follows (Table 2):

Table 2. Comparison of experimental results (error rate)

Method \ Iterations	CNN + MLP	CNN + LRM	CNN + Softmax
1	88.65%	5.46%	7.94%
5	5.17%	1.80%	1.86%
10	3.25%	1.37%	1.44%
20	1.76%	1.09%	1.12%
50	1.23%	0.86%	0.93%
80	1.00%	0.85%	0.94%
100	1.02%	0.74%	0.82%

In this experiment, there are 60000 training samples, each of the 50 samples to update a weight parameter, then an iteration will update the weight 1200 times.

5 Conclusions

In this thesis, we mainly study the use of the classifier based on logistic regression in the convolutional neural network, and introduce two kinds of classifier models based on logistic regression. There are many factors that affect the classifier performance, such as whether to use the hidden layer, the number of hidden layer neurons, and the learning rate of the regular term and so on. We can improve the convergence speed and test results by increasing the depth of the convolutional neural network. In practical applications, the classifier is selected according to the classification of the data set. In a word, it is a relatively simple and good method to apply logistic regression to convolutional neural network.

Acknowledgments. This work is supported by the Natural Science Foundation of Jiangsu Province (No. BK20130209), and the Project Funded by China Postdoctoral Science Foundation (No. 2014M560460).

References

1. LeCun, Y., Bengio, Y.: Convolutional networks for images, speech, and time-series. In: Arbib, M.A., (ed.) The Handbook of Brain Theory and Neural Networks, MIT Press, Cambridge, MA, USA (1995)
2. Bishop, M.: Pattern Recognition and Machine Learning. Springer, Heidelberg (2006)
3. Krizhevsky, A., Sutskever, I., Hinton, G.E.: ImageNet classification with deep convolutional neural networks. In: Proceedings of NIPs, pp. 1106–1114 (2012)
4. Simonyan, K., Zisserman, A.: Very deep convolutional networks for large-scale image recognition. In: Eprint Arxiv (2014)
5. Bouvrie, J.: Notes on convolutional neural networks (2006)
6. O'Neil, M.: Neural network for recognition of handwritten digits
7. LeCun, Y., Bottou, L., Bengio, Y., Haffner, P.: Gradient-based learning applied to document recognition. Proc. IEEE **86**, 2278–2324 (1998)
8. Huang, F.J., LeCun, Y.: Large-scale learning with SVM and convolutional for generic object categorization. In: Proceedings 2006 IEEE Computer Society Conference on Computer Vision and Pattern Recognition, vol. 1, pp. 284–291 (2006)
9. Vittinghoff, E., Glidden, D.V., Shiboski, S.C.: Regression methods in biostatistics: linear, logistic, survival, and repeated measures models. Springer, Heidelberg (2005)
10. Dreiseitl, S., Ohno-Machado, L.: Logistic regression and artificial neural network classification models: a methodology review. J. Biomed. Inform. **35**(5–6), 352–359 (2002)
11. Graves, A.: Supervised sequence labelling. In: Graves, A. (ed.) Supervised Sequence Labelling with Recurrent Neural Networks. SCI, vol. 385, pp. 5–13. Springer, Heidelberg (2012)
12. LeCun, Y., Boser, B., Denker, J., Henderson, D., Howard, R., Hubbard, W., Jackel, L.: Backpropagation applied to handwritten zip code recognition. Neural Comput. **1**(4), 541–551 (1989)
13. LeCun, B., Denker, J., Henderson, D., Howard, R., Hubbard, W., Jackel, L.: Handwritten digit recognition with a back-propagation network. In: NIPS (1990)

Event Detection with Convolutional Neural Networks for Forensic Investigation

Bo Yang[1,2], Ning Li[1,2(✉)], Zhigang Lu[1,2], and Jianguo Jiang[1]

[1] Institute of Information Engineering, Chinese Academy of Sciences,
Beijing 100093, China
{yangbo32,lining6}@iie.ac.cn
[2] Beijing Key Laboratory of Network Security Technology,
Beijing 100093, China

Abstract. Traditional approaches rely on domain expertise to acquire complicated features. Meanwhile, existing Natural Language Processing (NLP) tools and techniques are not competent to extract information from digital artifacts collected for investigation. In this paper, we propose an improved framework based on a Convolutional neural network (CNN) to capture significant clues for event identification. The experiments show that our solution achieves excellent results.

Keywords: Event detection · Convolutional neural networks · Digital investigation

1 Introduction

In digital investigations, the investigator typically has to handle substantial digital artifacts for forensics analysis. Among them, most are in the form of unstructured textual data, such as emails, chat logs, etc. The investigator searches clues from these data in order to answer questions about what happened, who caused the events, when events occurred, where, with whom they communicated, and so on. A pervasive problem is the fact that unstructured data are recorded using natural language, which is hard to understand completely by computer. This problem impedes the automation of crucial incriminating information retrieval and information extraction processes when facing a mass of textual data. Although the investigator can rely on modern digital forensics tools, such as executing keyword searches, this is often a manual process [1]. Due to the fact that the same concept is typically expressed by different terms and language styles, using keyword searches is not always effective. The quality of an analysis usually varies with the investigative experience and the expertise of each investigator [2].

With the advancement of data mining and availability of the computational re-sources to improve algorithm performance, methods of text mining using natural language processing techniques have gradually become available. By means of text mining, it makes the investigator easy to conduct content analysis and extract clues from textual data. Furthermore, automated techniques avoid threats to the privacy issues. For the reason that digital artifacts can reveal interested events, such as illegal

Z. Shi et al. (Eds.): IIP 2016, IFIP AICT 486, pp. 97–107, 2016.
DOI: 10.1007/978-3-319-48390-0_11

activities at key times, communications between suspects and victims, and other clues to the investigation, obtaining valuable information from them is based on the event detection task, which involves identification of events from specific types in the artifacts. Unlike traditional event extraction task defined in Automatic Content Extraction (ACE) evaluation [3], the event detection task in forensics varies according to case type and the investigator focuses a few specific types in a specific investigation. For example, the investigator should be interested in contact events, movement events etc. other than business events in a case of murder investigation. Current research on event extraction does not apply to digital forensics.

There are two categories of features for textual analysis on event detection: lexical-level features and sentence-level features. Convolutional neural networks that model sequential data have been proved to capture important information at sentence-level [4]. However, at the lexical-level, the same word influenced by its contexts has different meaning. Consequently, it frequently makes classifiers confused and also causes the inefficiency of keywords searches. In this paper, we present an improved one-layer convolutional neural network, which we name Classification with Similarity Vector by CNN (CSV-CNN), to capture more significant clues for event identification. Associated with each specified event is an event trigger, which most clearly expresses an event occurrence. We establish an event trigger lookup table containing most representative terms for a specific event to obtain lexical-level features. Our method computes and averages cosine similarity between each word in a sentence and word in the trigger lookup table in order to obtain a similarity vector. On the other hand, given input sentences, the network uses convolutional layers to acquire distributed vector representations of inputs to learn sentence-level features. The proposed network learns a distributed vector representation and similarity vector for each event class.

2 Preliminary

In this section, we first discuss related work in event extraction, and then introduce challenges posed by textual evidence.

2.1 Related Work

Since 1987, a series of Message Understanding Conferences (MUC) which were initiated by DARPA showed considerable interest in event extraction and encouraged the development of new methods of in information extraction [5]. MUC participants required submit evaluations to compete in textual information tasks. These tasks covered a wide range of goals, such as extracting fleet operations, identifying terrorist activities, detecting joint ventures and leadership changes in business, etc. The Automatic Content Extraction [3] program addresses the same problems as the MUC program, which defines extraction tasks according to the target objects, such as entity, relation and event. ACE focuses on annotating 8 event types and 33 subtypes.

There are several instances of the implementation of text mining techniques for event extraction can be found in literature. Best et al. [6] make use of a combination of entity extraction and a machine-learning technique for pattern-based event extraction. Li et al. [7] proposed the crossing-entity inference to improve the traditional sentence-level event extraction task. Li et al. [8] elaborate on a unified structure for extraction of entity mentions, relations and events in ACE task. These methods achieve relatively high performance on the basis of suitable choices of features selection. However, these methods suffer from complicated feature engineering and errors from existing NLP toolkits.

Over the past few years, deep learning methods have achieved remarkable success in machine learning, especially in computer vision and speech processing tasks [9, 10]. More recently, methods of handling NLP tasks using deep learning techniques started to overtake traditional sparse, linear models [11]. Word2Vec propose by Mikolov et al. [12], which could construct representation of words into a dense, low dimensional vector under a large corpus, has drawn great interests and is widely used as pre-trained word vectors. Convolutional neural networks (CNN), originally invented for computer vision, have subsequently realized significant performance in sentence classification. Nguyen et al. [13] used CNN for event detection that automatically learned feature form pre-trained Word2Vec embeddings, position embeddings, and entity type embeddings, with promising results.

2.2 Challenges Posed by Textual Evidence

Because of the efficiency and the convenience, the internet and computers are being used by criminals to facilitate their offenses. Most types of collected digital textual evidences are short texts, such as emails, chat logs, blogs, and tweets. In the past, these textual data have been studied for mining digital evidences, but all of them faced common challenges posed by noisy characteristics. First, these texts are short and have limited context. For example, emails seldom contain more than 500 words, and tweets can have less than 140 characters. Thus, statistical methods such as topic modeling do not apply to discover the themes of texts for insufficient words [14]. Second, most of them are more informal in style. The discourse can be regarded as written speech or spoken writing [15]. People do not always observe the grammatical rules and tend to make spelling mistakes. This means traditional NLP techniques are seldom appropriate for recorded conversations [16]. Third, people tend to use shortened terms, characters or punctuation symbols to convey or express more meaning or inside feelings or moods [17]. For example, the word "thanks" can be written as "thx", and the emoticon ":D" represents feeling happy. However, these terms or symbols are not captured by general-purpose tokenisers and not recognized as single tokens.

3 Model Description

In this paper, we formalize the event detection problem as a multi-class classification problem. Given a sentence, we want to make a prediction whether it expresses some event in the pre-defined event set or not [8]? In Subsect. 3.1, we first introduce standard one-layer CNN for sentence classification [4]. We then propose our augmentations in Subsect. 3.2, which exploit two stages: lexical-level and sentence-level feature extraction. Figure 1 describes the architecture of the proposed CSV-CNN.

3.1 Basic CNN

In this model, we first convert a tokenized sentence to a sentence matrix. Each token in a sentence is mapped into a low-dimensional vector representation, whether it is a word or not. It may be a fixed dimensional feature vector initialized at random and updated during training, or an output from pre-trained word2vec. We denote the dimensionality of the word vectors by n. In order to perform convolution on a sentence via non-linear filters, we expand the context to the maximum sentence length s by padding shorter sentences with special tokens. It is useful because it allows us to efficiently treat our data as an "image". Thus, the dimensionality of the sentence matrix is $s \times n$. In this matrix, let each row vector denote the corresponding token so that we can retain the inherent sequential structure of a sentence. We then denote a sentence x of length n by $x = \{x_1, x_2, \ldots, x_n\}$, and the sentence matrix is represented as $A \in \mathbb{R}^{s \times n}$.

The main idea behind a convolution and pooling computation is to apply non-linear filters over each instantiation of a k-word sliding window over a sentence. A filter $w \in \mathbb{R}^{k \times n}$ converts a window of k words into a fixed dimensional vector that achieves new features of the words in the window. We vary the dimensionality k of the filter to acquire different filters or use multiple filters for the same k to learn complementary features. In order to obtain a feature c_i, the filter is applied on a sub-part of A:

$$c_i = f(w \cdot x_{i:i+k-1} + b)$$

Where $x_{i:i+k-1}$ refers to the concatenation of tokens $x_i, x_{i-1}, \ldots, x_{i+k-1}$, $b \in \mathbb{R}$ is bias term, and f is a non-linear activation function that is applied element-wise. We execute convolution operations for each possible sub-matrix of A, which is $\{x_{1:k}, x_{2:k+1}, \ldots, x_{n-k+1:n}\}$, to produce a feature map:

$$c = [c_1, c_2, \ldots, c_{n-k+1}]$$

Where $c \in \mathbb{R}^{n-k+1}$. A max-pooling function is thus applied to the feature map to obtain the most salient information. Multiple filters are implemented and the outputs are concatenated into a "top-level" feature vector. Finally, this feature vector is passed to a fully connected softmax layer for classification.

Weights can be regularized in two ways. One is Dropout, in which we set values in the weight vector with a portion p at random during forward-backpropagation, the other is $l2$ norm constraint, for which we set a threshold λ for weight vectors during training; when it exceeds the threshold, we rescale the vector accordingly.

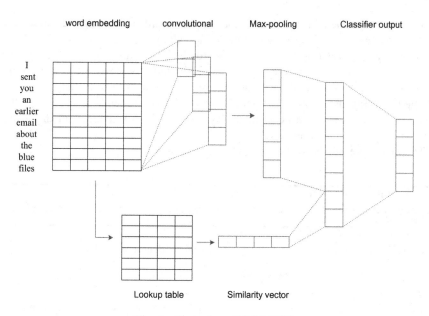

Fig. 1. Illustration of CSV-CNN

3.2   CSV-CNN

We propose an augmentation of CNN so that the model can learn more important features at the lexical-level. And the features at the sentence-level are left to learn by basic CNN model. We first select the most important words of each event type from data set. $TF - IDF$ [18] is one of the most popular algorithms used in the fields of in-formation retrieval and text mining. It provides a numerical statistic to reflect the extent of importance of a word in a collection. The occurrences of a word are directly proportional to the importance, but are inversely proportional to the frequency of the word in the collection. The $TF - IDF$ weight of word j is computed as follows:

$$TF - IDF\left(word_j\right) = TF\left(word_j\right) \times IDF\left(word_j\right)$$

Where TF represents the times a given word occurs in a specific document, IDF, which represents Inverse Document Frequency, is used to diminish the effect of words that appear too often in a collection. The IDF of $word_j$ is computed as follows:

$$IDF\left(word_j\right) = \log\frac{N}{\left(word_j\right)}$$

Where $DF\left(word_j\right)$ represents the quantity of documents containing $word_j$. We simply calculate a $TF\text{-}IDF$ score for each word and obtain the top n important words. These words can be seen as lexical-level features of each event type, which constitute the event trigger look up table.

We use pre-trained Word2Vec vectors for our word embeddings, exclude stop-words from sentences at the same time. Inspired by Word2Vec [12], which can automatically learn relationships between words like vec("king") – vec("man") = vec ("queen") – vec("woman"), we compute cosine similarity between the vectors of feature words from lookup table and the vectors of words from a sentence. We then average the results as a similarity vector corresponding to a specific event type. All of similarity vectors forms the penultimate layer of CSV-CNN along with the feature vector from the max-pooling layer.

The procedure of CSV-CNN is shown in the following.

Procedure CSV-CNN

Load_data()

Input: data set file obtained from the dataset folder

Output: x, y, vocabulary, vocabulary_inv, vocabulary_vector, similiar_vector

```
1.  for(int i1 = 1;i1<files.length;i1++)
2.      splits the data into words
3.          set the label, obtain keywords to each class(lookup table),
4.  end
5.  output(sentences, labels, keywords_labels)
6.  for(int i2 = 0;i2<number of sentences;i2++)
7.      Pads all sentences to the same length
8.  end
9.  Output(sentences_padded)
10. for(int i3 = 0;i3<number of words;i3++)
11.     Builds a vocabulary mapping from word to index
12.     Builds a vocabulary vector matrix mapping from word to vector
13. End
14. Output(vocabulary, vocabulary_inv, vocabulary_vector)
15. for(int i4=0;i4<number of sentence;i4++)
16.     for(int i5=0;i5<number of words in a sentence;i5++)
17.         Maps sentencs and labels to vectors based on a vocabulary
18.     end
19. end
20. Ouput(x,y)
21. for(int i6=0;i6<number of sentence;i6++)
22.     cleaning for stopwords of sentence, computes similarity for
        each sentence with keywords_labels
23. End
24. Ouput(similiar_vector)
```

CSV-CNN (*x, y, vocabulary, vocabulary_inv, vocabulary_vector, similiar_vector*)

```
1.   Initialize  CSV-CNN parameters;
2.   for(int i=0;i<number of input sentence;i++)
3.     for(int j=0;j<number of word of each sentence;j++)
4.          Builds a embedding mapping from words in each sentence
       to vector
5.       end
6.   end
7.   output(embedding);
8.   for(int k = 0; k<number of windows in the convolution; k++)
9.       Create a convolution + maxpool layer for each filter size
10.  end
11.  output(feature map)
12.  concatenate feature map with similiar_vector
13.  softmax and regularization
14.  output(predictions)
```

CSV-CNN training algorithm

```
1.   Initialize CSV-CNN parameters;
2.   for(int i=0;i< Number of training epochs;i++)
3.     for(training example x_batch, y_batch)
4.          calculates the cross-entropy loss for each class
5.          Backward compute_gradients
6.          Update_parameters(parameters, gradient)
7.       end
8.   end
9.   Output(parameters)
```

---

4   Experiments

4.1   Datasets

Although CNN has been demonstrated to have high performance in previous work on event detection [13], the dataset utilized for evaluation is based on newswire articles and does not apply to forensic investigation scenarios. Due to privacy constraints, actual cases and their related data are always not available for academic research. Therefore, we conducted a performance evaluation on a similar data. We utilized the Enron email dataset [19], which was made public during the legal investigation. This dataset contains messages belonging to 158 employees in Enron Corporation before its bankruptcy. We select sentences from these messages to tag events of interest. To this end, we establish four categories: movement, transaction, meet and correspondence. Each category contains 100 sentences. These four events usually occur in crime cases. Several example sentences are shown in Table 1. We performed experiments on one dataset containing 400 sentences with 1,271 unique terms (Fig. 2).

Table 1. Example sentence of dataset

Event type	Example sentence
Meet	We met with Gov Davis on Thursday evening in LA
Movement	I was at PIRA today and got a preview of their presentation on Oct 11–12
Correspondence	I sent you an earlier email about the blue files
Transaction	I transferred $10,000 out of the checking account on Monday 2/28/00

Table 2. Similar vector of example sentence

Meet	Movement	Correspondence	Transaction	Example sentence
0.08648066	0.07980033	0.04879327	0.02744178	We met with Gov Davis on Thursday evening in LA
0.0313917	0.0246127	0.04972452	0.0223587	I was at PIRA today and got a preview of their presentation on Oct 11–12
0.03896217	0.10931941	**0.28144109**	0.08396807	I sent you an earlier email about the blue files
0.0236352	0.05342571	0.07086471	0.04953432	I transferred $10,000 out of the checking account on Monday 2/28/00

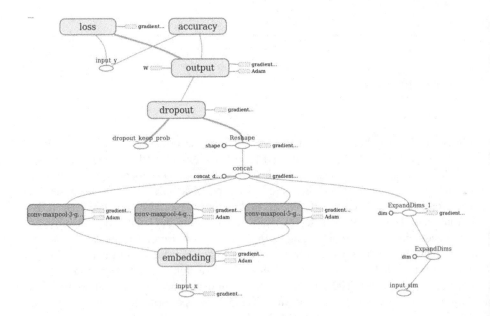

Fig. 2. Architecture of CSV-CNN

4.2   The Performance and Analysis

First, it demonstrates similar vector of our example sentences from our dataset as shown in Table 2. Although the similar vectors do not classify all of the example sentences independently, the first and the third ones can be classified by them. The result from Table 2 has proved the similar vector helps to find clues at the lexical-level indirectly.

The performance of CSV-CNN and standard CNN are shown in Figs. 3 and 4 respectively. Training metrics from the figures are not smooth because we use small batch sizes. It suggests that to achieve better results require a large corpus in the future work. As we can see from the figures, our model shows better at the accuracy and loss aspects.

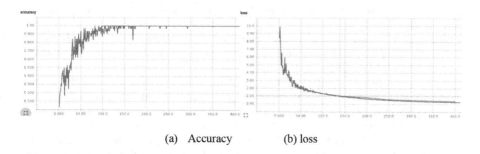

(a)   Accuracy   (b) loss

Fig. 3. Accuracy and loss plots of CSV-CNN (blue is training data, red is 10 % dev data). (Color figure online)

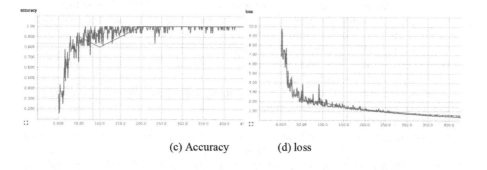

(c) Accuracy   (d) loss

Fig. 4. Accuracy and loss plots of CNN (blue is training data, red is 10 % dev data). (Color figure online)

5   Conclusions

In this paper, we present an improving framework based on CNN. We use prede-fined event typed sentences from enron dataset to evaluate the performance. The experiment shows that our solution achieves excellent results. In future work, we tend to refine our method in extracting specific information in wider scale. We plan on testing our method over a more appropriate large corpus to evaluate.

References

1. Pollitt, M.: A history of digital forensics. In: Chow, K.P., Shenoi, S. (eds.) Advances in Digital Forensics VI. IFIP Advances in Information and Communication Technology, pp. 3–15. Springer, Heidelberg (2010)
2. Al-Zaidy, R., Fung, B.C.M., Youssef, A.M., et al.: Mining criminal networks from unstructured text documents. Digital Invest. **8**(3), 147–160 (2012)
3. ADC. https://www.ldc.upenn.edu/collaborations/past-projects/ace
4. Kim, Y.: Convolutional neural networks for sentence classification (2014). arXiv preprint arXiv:1408.5882
5. Grishman, R., Sundheim, B.: Message understanding conference - 6: a brief history. In: Proceedings of the 16th International Conference on Computational Linguistics (COLING), vol. 1, pp. 466–471. Kopenhagen (1996)
6. Best, C., Piskorski, J., Pouliquen, B., et al.: Automating event extraction for the security domain. In: Chen, H., Yang, C.C. (eds.) Intelligence and Security Informatics: Techniques and Applications. Studies in Computational Intelligence, vol. 135, pp. 17–43. Springer, Heidelberg (2008)
7. Hong, Y., Zhang, J., Ma, B., et al.: Using cross-entity inference to improve event extraction. In: Proceedings of the 49th Annual Meeting of the Association for Computational Linguistics: Human Language Technologies, vol. 1, pp. 1127–1136. Association for Computational Linguistics (2011)
8. Li, Q., Ji, H., Hong, Y., et al.: Constructing information networks using one single model. In: EMNLP, pp. 1846–1851 (2014)
9. LeCun, Y., Bengio, Y., Hinton, G.: Deep learning. Nature **521**(7553), 436–444 (2015)
10. Graves, A., Mohamed, A., Hinton, G.: Speech recognition with deep recurrent neural networks. In: 2013 IEEE International Conference on Acoustics, Speech and Signal Processing (ICASSP), pp. 6645–6649. IEEE (2013)
11. Goldberg, Y.: A primer on neural network models for natural language processing (2015). arXiv preprint arXiv:1510.00726
12. Mikolov, T., Sutskever, I., Chen, K., et al.: Distributed representations of words and phrases and their compositionality. In: Advances in Neural Information Processing Systems, pp. 3111–3119 (2013)
13. Nguyen, T.H., Grishman, R.: Event detection and domain adaptation with convolutional neural networks, vol. 2, p. 365, Short Papers (2015)
14. Hua, W., Wang, Z., Wang, H., et al.: Short text understanding through lexical-semantic analysis. In: 2015 IEEE 31st International Conference on Data Engineering (ICDE), pp. 495–506. IEEE (2015)
15. Kucukyilmaz, T., Cambazoglu, B.B., Aykanat, C., et al.: Chat mining: predicting user and message attributes in computer-mediated communication. Inf. Process. Manage. **44**(4), 1448–1466 (2008)
16. Agarwal, S., Godbole, S., Punjani, D., et al.: How much noise is too much: a study in automatic text classification. In: Seventh IEEE International Conference on Data Mining (ICDM 2007), pp. 3–12. IEEE (2007)

17. Walther, J.B., D'Addario, K.P.: The impacts of emoticons on message interpretation in computer-mediated communication. Soc. Sci. Comput. Rev. **19**(3), 324–347 (2001)
18. Salton, G., McGill, M.J.: Introduction to Modern Information Retrieval. McGraw-Hill, New York (1986)
19. Cohen, W.W.: Enron email dataset. http://www.cs.cmu.edu/~enron/. Accessed 21 Aug 2009
20. Bach, J.: Modeling motivation in microPsi 2. In: Bieger, J., Goertzel, B., Potapov, A. (eds.) AGI 2015. LNCS, vol. 9205, pp. 3–13. Springer, Heidelberg (2015)
21. Abadi, M., Agarwal, A., Barham, P., et al.: Tensorflow: large-scale machine learning on heterogeneous distributed systems (2016). arXiv preprint arXiv:1603.04467

Boltzmann Machine and its Applications in Image Recognition

Shifei Ding[1,2(✉)], Jian Zhang[1,2], Nan Zhang[1,2], and Yanlu Hou[1,2]

[1] School of Computer Science and Technology,
China University of Mining and Technology, Xuzhou 221116, China
dingsf@cumt.edu.cn
[2] Key Laboratory of Intelligent Information Processing,
Institute of Computing Technology, Chinese Academy of Sciences,
Beijing 100190, China

Abstract. The overfitting problems commonly exist in neural networks and RBM models. In order to alleviate the overfitting problem, lots of research has been done. This paper built Weight uncertainty RBM model based on maximum likelihood estimation. And in the experimental section, this paper verified the effectiveness of the Weight uncertainty Deep Belief Network and the Weight uncertainty Deep Boltzmann Machine. In order to improve the images recognition ability, we introduce the spike-and-slab RBM (ssRBM) to our Weight uncertainty RBM and then build the Weight uncertainty spike-and-slab Deep Boltzmann Machine (wssDBM). The experiments showed that, the Weight uncertainty RBM, Weight uncertainty DBN and Weight uncertainty DBM were effective compared with the dropout method. At last, we validate the effectiveness of wssDBM in experimental section.

Keywords: RBM · DBM · DBN · Weight uncertainty

1   Introduction

The RBM is an unsupervised learning model which produces another expression of input data [1]. There are lots of training algorithms for RBM, such as Contrastive Divergence algorithm (CD), Persistent Markov chains and Mean Field methods, etc. In order to make full use of the features that extracted by RBM, Hinton et al. built the DBN model [2–4]. The DBN model provides a feasible method to train Multilayer Perceptron by the process of unsupervised pre-training. Another classic model in deep learning field is the Deep Boltzmann Machine (DBM). DBM is powerful in image recognition and image reconstruction [5]. And there are many other powerful models in deep learning field [6–8]. The Extreme Learning Machine (ELM) and Multilayer Extreme Learning Machine performed well in classification problem [9]. In the field of image recognition, lots of research has been done as well [10–12].

Overfitting is a common problem in neural networks. To address this question, lots of algorithms are proposed. Dropout method is used to alleviate the overfitting problem, which can be used in training RBM as well [13]. However, according to our experiments, the Dropout RBM is not good at image reconstruction, although it is

© IFIP International Federation for Information Processing 2016
Published by Springer International Publishing AG 2016. All Rights Reserved
Z. Shi et al. (Eds.): IIP 2016, IFIP AICT 486, pp. 108–118, 2016.
DOI: 10.1007/978-3-319-48390-0_12

powerful in image recognition. The Weight uncertainty method is also widely used in neural networks to alleviate the overfitting problems [14]. In this paper, the weight random variables are used in training RBM to alleviate the overfitting problems. In our experimental part, we validate the learning ability of Weight uncertainty RBM model. In classic RBM models, the conditional probabilities of visible units are binary. In Gaussian-binary RBM (mRBM) [15], the conditional probabilities of visible units follow Gaussian distribution. However, the mRBM performs not well in modeling nature images. In order to improve the images recognition ability, we introduce the spike-and-slab RBM (ssRBM) to our Weight uncertainty RBM and then build the Weight uncertainty spike-and-slab deep Boltzmann machine (wssDBM). At last, we validate the effectiveness of wssDBM in experimental section.

2 Restricted Boltzmann Machine and Semi-restricted Boltzmann Machine

2.1 Restricted Boltzmann Machine Models

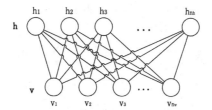

Fig. 1. The structure of RBM

RBM is a model based on energy functions. The structure of RBM is shown as Fig. 1:

The RBM model consists of a visible layer and a hidden layer. If the visible units and the hidden units are binary, the energy function can be defined as follow:

$$E\left(\vec{v}, \vec{h}\right) = -\sum_{i=1}^{n_v} a_i v_i - \sum_{j=1}^{n_h} b_j h_j - \sum_{i=1}^{n_v}\sum_{j=1}^{n_h} h_j \times w_{ji} \times v_i \qquad (1)$$

$\vec{a}$ is the bias vector of the visible layer, $\vec{b}$ is the bias vector of the hidden layer, W is the weight matrix between visible units and hidden units, $\vec{v}$ is the visible layer vector, $\vec{h}$ is the hidden layer vector. Then, the probability based on $E\left(\vec{v}, \vec{h}\right)$ is shown as formula 2:

$$P\left(\vec{v}, \vec{h}\right) = \frac{1}{Z} e^{-E\left(\vec{v}, \vec{h}\right)} \qquad (2)$$

Z is the partition function.

$$Z = \sum_{v,h} e^{-E(\vec{v},\vec{h})} \tag{3}$$

According to the whole training set, the likelihood function is defined as:

$$L_s = \ln \prod_{i=1}^{n_s} P(\vec{v^i}) = \sum_{i=1}^{n_s} \ln P(\vec{v^i}) \tag{4}$$

n_s is the number of samples. And there are many algorithms can be used to maximize the likelihood function, such as Stochastic Gradient Descent algorithm. Let $\theta = (\vec{a}, \vec{b}, W)$, the derivative of the likelihood function is shown as formula 5:

$$\frac{\partial L_s}{\partial \theta} = -\sum_{i=1}^{n_s} \left(\sum_h P(\vec{h}|V^{(i)}) \frac{\partial E(V^{(i)}, \vec{h})}{\partial \theta} + \sum_{v,h} P(\vec{v}, \vec{h}) \frac{\partial E(\vec{v}, \vec{h})}{\partial \theta} \right) \tag{5}$$

θ is the parameter. And the conditional probabilities are shown as follows:

$$p(h_k = 1|\vec{v}) = sigmoid\left(b_k + \sum_{i=1}^{n_v} w_{ki} v_i \right) \tag{6}$$

$$p\left(v_k = 1|\vec{h} \right) = sigmoid\left(a_k + \sum_{j=1}^{n_v} h_j w_{kj} \right) \tag{7}$$

Hinton et al. proposed Contrastive Divergence (CD) algorithm to approximate the Maximum Likelihood Estimation. Based on single sample, and k is the number of steps in K-steps Contrastive Divergence algorithm (CD-K). We update the weights between visible units and hidden units with the following formulas:

$$\Delta w_{ij} = \eta_w \left(P\left(h_i = 1|\vec{v}^{(0)} \right) \vec{v}^{(0)} - P\left(h_i = 1||\vec{v}^{(k)} \right) \vec{v}^{(k)} \right) \tag{8}$$

$$\Delta a_i = \eta_a \left(v_i^{(0)} - v_i^{(k)} \right) \tag{9}$$

$$\Delta b_i = \frac{\partial \ln P(\vec{v})}{\partial b_i} \approx \eta_b \left(P\left(h_i = 1|\vec{v}^{(0)} \right) - P\left(h_i = 1|\vec{v}^{(k)} \right) \right) \tag{10}$$

η is the learning rate.

2.2 Spike-and-Slab Restricted Boltzmann Machine

In order to model the expectation and covariance of Gaussian distribution, ssRBM model is proposed. In ssRBM, a variable slab is used to express the density. Based on the variable slab, the conditional probability of visible units has a diagonal covariance matrix. And the block Gibbs sampling can be used in ssRBM. The energy function can be expressed as follow:

$$E(v, s, h) = \frac{1}{2} v^T \Lambda v - \sum_{i=1}^{N} \left(v^T W_i s_i h_i + \frac{1}{2} s_i^T \alpha_i s_i + b_i h_i \right) \tag{11}$$

Beyond this energy function, the conditional probability can be expressed as follows:

$$p(v|h) = \frac{1}{B} N \left(0, \left(\Lambda - \sum_{i=1}^{N} h_i W_i \alpha_i^{-1} W_i^T \right)^{-1} \right) \tag{12}$$

$$p(s|v, h) = \prod_{i=1}^{N} N \left(h_i \alpha_i^{-1} W_i^T v, \alpha_i^{-1} \right) \tag{13}$$

$$P(h_i = 1|v) = sigmoid \left(\frac{1}{2} v^T W_i \alpha_i^{-1} W_i^T v + b_i \right) \tag{14}$$

$$p(v|s, h) = N \left(\Lambda^{-1} \sum_{i=1}^{N} W_i s_i h_i, \Lambda^{-1} \right) \tag{15}$$

In this way, the conditional probability of visible units is a diagonal covariance matrix.

3 The Training Algorithms About RBM and Boltzmann Machine

There are lots of training algorithms for RBM model. Early, Persistent Markov chains and the Simulated Annealing method were used to estimate the data independent expectation and data dependent expectation. Although CD algorithm is not accurate in learning step-size, it guarantees the correct gradient direction. Based on CD algorithm, Persistent Contrastive Divergence algorithm (PCD) and Persistent Contrastive Divergence algorithm with Fast weights (FPCD) are proposed. In order to decrease the sampling time in training process, Mean Field Method is proposed.

3.1   Mean Field Method

The detailed Mean Field Method is shown in reference. In the probabilistic graphical models, the real posterior distribution $P(h|v; \theta)$ is replaced by an approximate posterior distribution $Q(h|v; \lambda)$. What we need to do is minimizing the following *KL* Divergence:

$$\lambda^* = \arg \min_{\lambda} KL[Q(h|v)||P(h|v)] \tag{16}$$

For the Mean Filed Boltzmann Machine, we have:

$$Q(h|v, u) = \prod_{i \in H} u^{S_i}(1 - u_i)^{1-S_i} \tag{17}$$

Then, the KL Divergence can be expressed as follow:

$$KL[Q||P] = \sum_i (u_i \ln u_i + (1 - u_i) \ln(1 - u_i)) - \sum_{i<j} \theta_{ij} u_i u_j - \sum_i \theta_i^c u_i + \ln Z_c \tag{18}$$

Z_c is the partition function, $Z_c = \sum_{\{H\}} \left(\exp \left(\sum_{i<j} \theta_{ij} S_i S_j + \sum_i \theta_i^c S_i \right) \right)$, $\theta_i^c = \theta_i + \sum_{j \in V} \theta_{ij} S_j$

S_i and S_j are independent variables. The expectation of S_i is u_i, and the expectation of S_j is u_j.

$$u_i = sigmoid \left(\sum_j \theta_{ij} u_j + \theta_i \right) \tag{19}$$

In general, the EM algorithm can be used in the Mean Field inference. Evidence shows that, for the same test data, the Mean Field Method is 10 to 30 times faster than Gibbs sampling. The Mean Field Method can be used in RBM model:

$$\ln P(v; \theta) \geq \sum_{i,j} W_{ij} v_i u_j + \sum_i b_i v_i - \ln Z - \sum_j \left(u_j \ln u_j + (1 - u_j) \ln(1 - u_j) \right) \tag{20}$$

The probability values of the hidden unit can be expressed as:

$$u_i = sigmoid \left(\sum_j W_{ij} v_j + b_i \right) \tag{21}$$

3.2   Persistent Markov Chain

The detailed Persistent Markov chain algorithm is shown in reference [16, 17]. If the Markov chain is long, and the step-size is not too large, the Markov chain will reach the steady state. The Persistent Markov chains can be used in training Boltzmann Machines as well. For the data independence expectation, we can obtain an effective approximation. The algorithm of Persistent Markov chains is shown in Table 1:

Table 1. Algorithm of Persistent Markov chains

Algorithm of Persistent Markov chains
Randomly initialize θ_0 and M sample particles $\left\{ \tilde{x}^{0,1}, ..., \tilde{x}^{0.M} \right\}$.
for $t = 0 : T$ (number of iterations) do   for $i = 1 : M$ (number of Parallel Markov chains) do
Sample $\tilde{x}^{t+1,i}$ given $\tilde{x}^{t,i}$ using transition operator $T_{\theta^t}\left(\tilde{x}^{t+1,i} \leftarrow \tilde{x}^{t,i} \right)$.
end for
Update: $\theta^{t+1} = \theta^t + \alpha_t \left[\Phi\left(\overline{x} \right) - \dfrac{1}{M} \sum_{m=1}^{M} \Phi\left(\tilde{x}^{t+1,m} \right) \right]$
Decrease α_t.
End for.

θ is a set of parameters, Φ is sufficient statistics vector, α_t is the learning rate.

4 Deep Belief Networks and Deep Boltzmann Machine

4.1 Deep Belief Networks

DBN is a hybrid network, which is proposed by Hinton in 2006. The top 2 layers consist of an associative memory with undirected connections. And the layers below have directed, top-down generative connections. In training process of DBN, the network is initialized layer by layer. Suppose that DBN is a model which has infinite layers. Then we use the same weight W_0 to initialize the network, the model can be considered as RBM in the training process, which is shown in Fig. 2(a). After training the first layer of DBN, the weights of the first layer remain constant, and the other weights are replaced by W_1. In this case, the priori information will be updated layer by layer. Hinton et al. proved that the pre-training process can tighten the variable boundary: $\ln p(v|W_1, W_2) \geq \ln p(v|W_1)$, and the pre-training process of DBN is shown as Fig. 2(b):

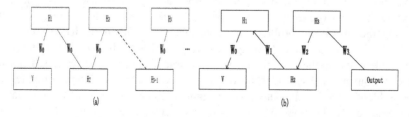

(a)   (b)

Fig. 2. Shows the diagram of training process in DBN.

After pre-training in DBN, the DBN model can be fine-tuned by BP algorithm as a neural network.

4.2   Deep Boltzmann Machine

Different from DBN, DBM is still a Boltzmann Machine in topology. In the training process of DBM, the activation of each unit depends on the units in the adjacent layers. Salakhutdinov pointed out that the training process of DBM model can also be carried out layer by layer. However, different from DBN, Salakhutdinov showed that the different effects can be obtained by replacing the priori information with different proportions.

The probabilities in DBM model can be expressed as follows:

$$p\left(h_j^1 = 1 \middle| \vec{v}, \vec{h}^2\right) = sigmoid\left(\sum_i W_{ij}^1 v_i + \sum_m W_{jm}^2 h_m^2 + b_j^1\right) \tag{22}$$

$$p\left(h_m^2 = 1 \middle| \vec{h}^1, \vec{h}^3\right) = sigmoid\left(\sum_j W_{jm}^2 h_j^1 + \sum_l W_{ml}^3 h_l^3 + b_m^2\right) \tag{23}$$

$$p\left(h_l^3 = 1 \middle| \vec{h}^2\right) = sigmoid\left(\sum_m W_{ml}^3 h_m^2 + b_l^3\right) \tag{24}$$

$$p\left(v_i = 1 \middle| \vec{h}^1\right) = sigmoid\left(\sum_j W_{ij}^1 h_j^1 + b_j\right) \tag{25}$$

The superscripts represent the layer number. The log-likelihood can be approximated by using Stochastic Approximation algorithm and Mean Field Algorithm.

5   Weight Uncertainty Spike-and-Slab Restricted Boltzmann Machine

5.1   Weight Uncertainty Method

In the whole training process, the weights and biases need to be calculated. And the weights and biases are regarded as real valued variables. In this case, training a neural network prefer to encounter the problem of overfitting. There are lots of research about alleviating the overfitting problem in neural networks. Based on RBMs, the main algorithm is dropout method. Although dropout RBM is useful to alleviate the over-fitting problem in classification, the image reconstruction ability of dropout RBM is not better than conventional RBM. If the weights are considered as random variables, the above problems may be alleviated. The weights are considered as random variables, and we assume that the random variables follow Gaussian distribution. What we need to do is calculating the expectation and the covariance. And the generations of different weights can be regarded as the sampling from Gaussian distribution. Therefore, the Weight uncertainty neural network can be considered as the ensemble of neural networks.

In the research of Blundell et al., all weights in networks are regarded as probability distributions, rather than a real value. The objective is to find a variational approximation to the Bayesian posterior distribution on the weights. And the objective function can be expressed as follows:

$$
\begin{aligned}
\theta' &= \arg\min_{\theta} KL[q(w|\theta)||P(w|\theta)] \\
&= \arg\min_{\theta} KL[q(w|\theta)||P(w)] - E_{q(w|\theta)}[\log P(D|w)]
\end{aligned}
\tag{26}
$$

According to the thought of MAP estimation, let

$$
f(w,\theta) = \log q(w|\theta) - \log P(w)P(D|w)
\tag{27}
$$

In RBM model, in order to obtain more effective image reconstruction and classification, we introduce the weight random variables to RBM model. The cost functions of RBM can be written as: maximum likelihood estimation $p(v|w)$, and MAP estimation $p(w|v)$. In order to simplify calculation, we use the Maximum Likelihood Estimation to calculate the activation probabilities. Assuming the weight W follows the Gaussian distribution, the mean value of W is μ, standard deviations are $\sigma = \log(1 + \exp(\rho))$, if $\varepsilon \sim N(0, I)$, the weights can be expressed as: $w = \mu + \log(1 + \exp(\rho)) \circ \varepsilon$. According to the chain rule, the derivatives can be expressed as follows:

$$
\frac{\partial \log p(w_{ij})}{\partial w_{ij}} \times \frac{\partial w_{ij}}{\partial \mu} = \left(P(h_j = 1|\vec{v})v_i - \sum_{v,h} P(v)P(h_j = 1|\vec{v})v_i \right) \times 1
\tag{28}
$$

$$
\frac{\partial \log p(w)}{\partial w} \times \frac{\partial w}{\partial \rho} = \left(P(h = 1|\vec{v})v - \sum P(v)P(h = 1|\vec{v})v \right) \times \frac{\varepsilon}{1 + \exp(-\rho)}
\tag{29}
$$

In the experimental section, we test the classification ability and image reconstruction ability of Weight uncertainty RBM model (WRBM), and then build the DBN and DBM based on WRBM.

5.2   Weight Uncertainty Spike-and-Slab Deep Boltzmann Machine

ssRBM is used to model nature images. In this paper, we use the ssRBM as the feature extractor, and build the DBM model, and then we introduce the weight random variables to the DBM, and build the wssDBM. At last, we validate the effectiveness of wssDBM in experimental section.

6 Experimental Analysis

Firstly we compare the WRBM with RBM and dropout RBM in classification and image reconstruction. The algorithm we used in fine-tuning process is the conjugate gradient algorithm, the iterative steps are 100. In this experiment, we use MNIST, MNIST-Basic and Rectangles as the testing data sets. The attributes of these data sets are shown in Table 2:

Table 2. The attributes of data sets

	Number of training samples	Number of testing samples	Attributes	Labels
MNIST-Basic	10000	50000	784	10
Rectangles	1000	50000	784	2
MNIST	60000	10000	784	10

Firstly we test the image recognition ability of the WRBM (WRBM). In fine-tuning process. The testing accuracies are shown in Table 3:

Table 3. The number of misclassifications in shallow models

	MNIST-Basic	Rectangles
RBM-BP	1811	2586
Dropout RBM-BP	1633	2175
WRBM-BP	1567	1979

As we can see from Table 3, the classification accuracies of WRBM are better than RBM and dropout RBM, that is to say, like dropout method, the weight random variables are useful in classification problems.

The reconstruction errors in training process are shown in Table 4:

Table 4. The reconstruction errors of RBM and WRBM

	MNIST-Basic
RBM	61631
Dropout RBM	65623
WRBM	52638

As we can see from Table 4, the image reconstruction ability of WRBM is better than other models. And the weight random variables are also useful in image reconstruction.

The topologies in DBM and Weight uncertainty DBM (WDBM) are 784-1000-1000-10. And the topologies in DBN and Weight uncertainty DBN (WDBN)

Table 5. The number of misclassifications of DBN and DBM

	MNIST-Basic	MNIST	Rectangles
DBM	1115	94	1309
WDBM	1016	92	1139
DBN	1283	105	1278
WDBN	1251	101	367
Dropout DBN	1257	99	778
wssDBM	1022	103	477

are 784-1000-2000-10. The iterative steps in RBM training process are 200. The iterative steps in DBM training process is 300. The testing accuracies are shown in Table 5.

As we can see from Table 5, the WDBM performs better than conventional DBM model, and WDBN is also comparable to dropout DBN in classification problem. wssDBM also performs well in classification problems.

7 Conclusion

In this paper, in order to alleviate the overfitting problem, and improve the ability of image reconstruction in RBM model, we introduce the Weight uncertainty method to RBM. The WRBM performs well in our experiments. In our experiments, the Weight uncertainty method is useful in both classification and image reconstruction. Intuitively speaking, the weight random variables can be regarded as the ensemble of neural networks. And the wssDBM is useful in image recognition.

Acknowledgements. This work is supported by the National Natural Science Foundation of China (Nos. 61379101, 61672522), and the National Key Basic Research Program of China (No. 2013CB329502).

References

1. Hinton, G.E.: Training products of experts by minimizing contrastive divergence. Neural Comput. **14**(8), 1771–1800 (2002)
2. Roux, N., Bengio, Y.: Representational power of restricted Boltzmann machines and deep belief networks. Neural Comput. **20**(6), 1631–1649 (2008)
3. Hinton, G.E., Osindero, S., Teh, Y.: A fast learning algorithm for deep belief nets. Neural Comput. **18**(7), 1527–1554 (2006)
4. Hinton, G.E., Salakhutdinov, R.: Reducing the dimensionality of data with neural networks. Science **313**(5786), 504–507 (2006)
5. Lee, H., Pham, P.T., Yan, L., et al.: Unsupervised feature learning for audio classification using convolutional deep belief networks. In: Advances in Neural Information Processing Systems, pp. 1096–1104 (2009)

6. Norouzi, M., Ranjbar, M., Mori, G.: Stacks of convolutional Restricted Boltzmann Machines for shift-invariant feature learning. In: Computer Vision and Pattern Recognition, pp. 2735–2742 (2009)
7. Salakhutdinov, R., Larochelle, H.: Efficient learning of deep Boltzmann Machines. J. Mach. Learn. Res. 9(8), 693–700 (2010)
8. Salakhutdinov, R., Hinton, G.E.: An efficient learning procedure for deep Boltzmann machines. Neural Comput. 24(8), 1967–2006 (2012)
9. Zhang, J., Ding, S.F., Zhang, N., et al.: Incremental extreme learning machine based on deep feature embedded. Int. J. Mach. Learn. Cybern. (2015, to be publised)
10. Zhang, N., Ding, S.F., Shi, Z.Z.: Denoising Laplacian multi-layer extreme learning machine. Neurocomputing 171, 1066–1074 (2016)
11. Ding, S.F., Zhang, N., Xu, X.Z., et al.: Deep extreme learning machine and its application in EEG classification. Mathematical Problems in Engineering, pp. 1–11 (2015)
12. Zheng, Y., Jeon, B., Xu, D., M, Q., et al.: Image segmentation by generalized hierarchical fuzzy C-means algorithm. J. Intell. Fuzzy Syst. 28(2), 961–973 (2015)
13. Srivastava, N., Hinton, G.E., Krizhevsky, A.: Dropout: a simple way to prevent neural networks from overfitting. J. Mach. Learn. Res. 15, 1929–1958 (2014)
14. Blundell, C., Cornebise, J., Kavukcuoglu, K.: Weight uncertainty in neural networks. In: Proceedings of the 32nd International Conference on Machine Learning, Lille, France (2015)
15. Krizhevsky, A., Hinton, G.E.: Learning multiple layers of features from tiny images, Technical report, U. Toronto (2009)
16. Tieleman, T.: Training restricted Boltzmann machines using approximations to the likelihood gradient. In: Proceedings of the 25th International Conference on Machine Learning, pp. 1064–1071. ACM (2008)
17. Tieleman, T., Hinton, G.E.: Using fast weights to improve persistent contrastive divergence. In: Proceedings of the 26th International Conference on Machine Learning, pp. 1033–1040. ACM (2009)

Social Computing

Trajectory Pattern Identification and Anomaly Detection of Pedestrian Flows Based on Visual Clustering

Li Li[(✉)] and Christopher Leckie

Department of Computing and Information Systems, The University of Melbourne,
Parkville, Melbourne, VIC 3010, Australia
lli10@student.unimelb.edu.au, caleckie@unimelb.edu.au

Abstract. Extracting pedestrian movement patterns and determining anomalous regions/time periods is a major challenge in data mining of massive trajectory datasets. In this paper, we apply contour map and visual clustering algorithms to visually identify and analyse areas/time periods with anomalous distributions of pedestrian flows. Contour maps are adopted as the visualization method of the origin-destination flow matrix to describe the distribution of pedestrian movement in terms of entry/exit areas. By transforming the origin-destination flow matrix into a dissimilarity matrix, the iVAT visual clustering algorithm is applied to visually cluster the most popular and related areas. A novel method based on the iVAT algorithm is proposed to detect normal/abnormal time periods with similar/anomalous pedestrian flow patterns. Synthetic and large, real-life datasets are used to validate the effectiveness of our proposed algorithms.

Keywords: Data mining · Pedestrian trajectory pattern · Visualization · Clustering · iVAT algorithm

1 Introduction

There is growing interest in the problem of extracting useful information from massive trajectory datasets derived by various sensing methods. Understanding patterns of pedestrian movement is useful in applications such as pedestrian flow management, public security and safety. A major challenge in pattern analysis of pedestrian movement is how to discover and describe the movement patterns hidden in trajectories, and identify any misbehaviour or interesting events.

The main approaches to trajectory data analysis and anomaly detection fall into the category of trajectory data mining. To detect and recognize social events, two common approaches used to address this problem are statistical methods combined with classification and clustering-based methods. Many existing approaches to address this problem have the limitations that they focus on the

© IFIP International Federation for Information Processing 2016
Published by Springer International Publishing AG 2016. All Rights Reserved
Z. Shi et al. (Eds.): IIP 2016, IFIP AICT 486, pp. 121–131, 2016.
DOI: 10.1007/978-3-319-48390-0_13

details of trajectories, but do not consider the characteristics of the trajectory distribution.

In this paper, we address this limitation of existing approaches by proposing the use of contour maps and visual clustering. Contour maps are a very useful visualization tool for three-dimensional data, which we adopt to visually describe the connection between different subareas and describe the distribution of trajectories. Visual clustering methods such as VAT (Visual Assessment of cluster Tendency) and iVAT (improved Visual Assessment of cluster Tendency) [1,9] are proposed to visually assess the clustering tendency of a set of objects. By using the VAT/iVAT approach, we are able to visualize and determine the possible number of clusters of locations or the periods with similar activity, and then determine abnormal areas/days with significantly different trajectory distributions.

The main contributions of our paper are as follows. First, we use a visualization method to describe pedestrian movement distributions in terms of their origin and destination points. By transforming the origin-destination flow matrix into a dissimilarity matrix, we visually cluster the most popular flows using the VAT/iVAT algorithms. The popular flow patterns are important for monitoring the safety and security of public areas. Second, we propose a novel method of detecting abnormal time periods with anomalous pedestrian trajectory distributions based on the results of the VAT/iVAT algorithms. By doing so, it is possible to detect the occurrence and impact of special events. Finally, we evaluate our methods and make relevant comparisons on a large, real-life dataset, the Edinburgh informatics forum database [7], and demonstrate the effectiveness of our proposed algorithms.

2   Related Work

An important aspect of a monitoring system is to detect significant events or unusual behaviour in that environment. By mining pedestrian trajectories, it is possible to detect the occurrence of major events, such as celebrations, parades, business promotions, accidents, disasters and others, which may be threats to public security or safety. With the increasing availability of big trajectory data, there have been various research methods on the detection and recognition of anomalous social events from trajectory data. In this paper, we focus on the challenges of how to detect normal patterns of flows, and how to detect abnormal pedestrian flow patterns.

The first problem we address is to find and visualize the most visited subareas in a given region. In [3], an algorithm for extracting popular regions was proposed, with the popular regions defined as regions with trajectory densities larger than a given threshold, which is manually prescribed instead of adaptively adjusted. In this paper, we aim to find a parameter free algorithm for this problem. Liu et al. [4] designed real time analytical method for spatial-temporal data of daily travel patterns in metropolitan urban environments. The data for analysis of Liu [4] are taxi traces and smart card records, whereas in contrast we mainly focus on pedestrian trajectories. While they provided an analysis on travel patterns, they did not consider the problem of detecting anomalies. Liu et al. [5] and Lu et al. [6] shared the same idea of extracting entry/exit points and using length analysis to derive indoor scene structure and identify abnormal motion behaviours. However, the methods in [5,6] only detect particular instances of trajectories. In this paper, using the same dataset [5,6], we focus on using the origin-destination matrix as a whole to describe the pedestrian flow patterns and cluster the most visited areas.

The second problem we address is how to detect a set of time periods with abnormal trajectory motion patterns. In the literature, there have been many research studies on anomaly detection methods for traffic or pedestrian data. Pang et al. [8] proposed a statistical model, which adapts likelihood ratio tests to find anomalous regions for monitoring the emergence of unexpected behaviour based on GPS data from taxis. Chawla et al. [2] adopted Principal Component Analysis (PCA) to detect traffic anomalies from GPS data. In [2], moving activities of a crowd were simulated as the movements of a group of points, and the distribution of point groups is described with fractal dimensions. PCA was used to remove the disturbed factors from a feature vector and maintain only relevant information. Witayangkurn et al. [10] proposed a framework based on a hidden Markov model to construct a pattern of spatial-temporal movement of people in each area in a grid during each time period. However, they focused on using changes in the population to detect anomalies. In this paper, we intend to detect anomalous time periods of trajectory distributions based on the use of visual clustering methods, which can partition the data into clusters in a visual manner. In this way, we can provide a robust, unsupervised approach for clustering periods of normal pedestrian activity and visually highlighting periods of anomalous pedestrian activity.

3   Problem Statement

Our aim is to find the most popular routes of pedestrians in a given region and check which areas are the most related by visualizing the motion patterns, and to detect and visualize abnormal time periods by comparing the distributions of motion patterns. An outlier or anomaly in a dataset is considered to be an inconsistent observation (or subset of observations) compared with the remainder of that set of data, such as a substantial change in the popularity of pedestrian routes.

Suppose that the structure of the monitored area is known and can be divided into k subareas $A = \{A_1, A_2, , A_k\}$ according to the functions of different areas or some other predefined labelling. For example, as shown in Fig. 1 (from the Edinburgh informatics forum database), there are 13 functional areas in this forum.

The dataset consists of a set of detected people walking through these subareas. Suppose we are given the trajectory data $T = \{T_{day_1}, T_{day_2}, ..., T_{day_m}\}$ covering a set of m days $D = \{day_1, day_2, .., day_m\}$, where on day_i a set of trajectories

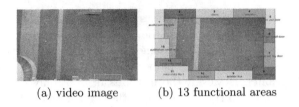

(a) video image   (b) 13 functional areas

Fig. 1. Video image and functional areas of the Edinburgh informatics forum

$T_{day_i} = \{T_{i1}, T_{i2}, ..., T_{in_i}\}$ has been collected, where n_i is the number of trajectories on day_i, and it is assumed that the trajectory data is the outcome of a robust tracking system. Each trajectory is composed of triplets (x, y, t) containing X/Y coordinates and sampling time $T_{ij} = \{(x_1, y_1, t_1), (x_2, y_2, t_2), .., (x_l, y_l, t_l)\}_{ij}$, where l is the number of points in trajectory T_{ij}. Given trajectory dataset T, our aim is to find and visualize the most visited subareas $A' \subset A$ and identify a set of time periods $D' \subset D$ with abnormal trajectory motion patterns.

To address this aim, we examine three related research questions: (1) how to identify and summarize related sets of pedestrian flows over a given time period; (2) how to identify which time periods exhibit similar patterns of pedestrian flows; and (3) how to identify which time periods have experienced anomalous flow patterns? We present our approach to these research questions in the following three sections.

4   Case Study - Edinburgh Pedestrian Flow

The trajectory dataset from the University of Edinburgh is used as a case study in this paper. The dataset provides the scene under surveillance and the configuration of it and the image covers most of the main hall, which has been shown in Fig. 1. The most significant features of the hall are that there are many entry and exit points, i.e., the main entrance to the building, lifts, access to the Atrium, access to the second part of the hall, staircase, reception desk, and the four other exits, which means that there are a variety of possible pedestrian flows in this area.

The dataset consists of a set of detected targets of people walking through the Informatics Forum, the main building of the School of Informatics at the University of Edinburgh. The valid data covers 118 days of observation, resulting in about 90,000 observed trajectories in total. Substantial differences are observed between weekdays (Mon-Fri) and weekends. The average number of trajectories on weekdays is 932, which is significantly larger than the number of 140 on weekends.

Some papers [5–7] provided clustering methods to exact the pedestrian flows from trajectory data in terms of these entry and exit points. Based on the pedestrian trajectories detected from video images in Majecka et al. [7], which are represented by a sequence of centroid positions, we provide a visual clustering using contour maps and iVAT.

5   Summarizing Related Flows

The first question we address is how to identify and summarize related sets of pedestrian flows over a given time period. The challenges in addressing this question are how to summarize a given pedestrian flow matrix so that a user can identify the dominant flows, and then how to identify related subsets or clusters of flows. In this way, we can summarize the pedestrian activity over a given period of time. In this section, we illustrate methods to summarize a flow matrix and detect normal/abnormal days by identifying related flows.

5.1   Synthetic Cases

The rows and columns of a contour map reflect the order in which entry/exit pairs are labelled, but it does not reflect the clustering relationships between flows, so we would like to visually group related flows. Consider the following synthetic example.

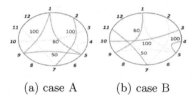

(a) case A   (b) case B

Fig. 2. Two synthetic cases. Italic numbers from 1 to 12 represent 12 different areas. Lines and dashed lines between different areas represent the large pedestrian flows and small pedestrian flows respectively, and numbers next to lines indicate the size of pedestrian flows (Color figure online)

As shown in Fig. 2, these two synthetic cases both have four relatively high pedestrian flows 50, 80, 100 and 100, but the relationships between flows are different. To visually display the origin-destination pair distribution characteristics, we introduce contours to represent the flow matrix. Contours indicate equal valued regions with the same colour. This is similar to a heat map, and the characteristics of the matrix distribution can be visually analysed. The application of contours helps us better visualize and compare distribution characteristics of the origin-destination flow matrix, which would otherwise be hard to analyse only by the values of the matrix. The contour maps of the two cases are shown in Fig. 3 respectively.

Although we can easily find the distribution of trajectories flows and detect origin/destination pairs that have high pedestrian flows in Fig. 3, we cannot easily identify related flows, since the light areas are scattered on the contour map. The main challenge is how to reorder the rows/columns of a contour map to group related flows. In this paper, we propose to treat this as a visual clustering problem.

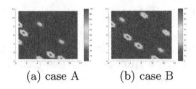

(a) case A   (b) case B

Fig. 3. Contour maps of the two synthetic cases in Fig. 2

VAT and iVAT are useful tools for visual assessment of clustering tendency, as is shown by Bezdek et al. [1] and Wang et al. [9]. The VAT algorithm displays a reordered dissimilarity matrix D as a grey-scale image with a modified version of Prims minimal spanning tree algorithm. The iVAT algorithm augments VAT by applying a path-based distance transform to the input dissimilarity data before VAT images are made. It reorders the dissimilarity matrix of the given set of objects so that it can display any clusters as dark blocks along the diagonal of the image, and a diagonal dark block appears in the iVAT image only when a tight group exists in the data. In this paper, we only provide the results based on the iVAT algorithm. The main steps of iVAT are:

Step 1: Transform input dissimilarity matrix $D \rightarrow D'$ using a path-based distance;

Step 2: VAT is applied to reorder $D' \rightarrow D'^*$, resulting in an iVAT image $I(D'^*)$ whose $(i,j)^{th}$ element is a scaled dissimilarity value between objects o_i and o_j.

Since a dissimilarity matrix D is the input data to the iVAT algorithm, a method to transform the origin-destination matrix F to a dissimilarity matrix D is proposed in our paper. Considering that the origin-destination flow matrix F is non-symmetric (F_{ij} and F_{ji} may be different), the first step is to transform the flow matrix to be symmetric. There are three methods to derive a symmetric matrix S: (1) $S_{ij} = S_{ji} = max(F_{ij}, F_{ji})$; (2)$S_{ij} = S_{ji} = min(F_{ij}, F_{ji})$; (3) $S_{ij} = S_{ji} = (F_{ij} + F_{ji})/2$.

This symmetric flow matrix S can be normalized by using $S'_{ij} = S_{ij}/S_{max}$, where S_{max} is the value of the largest element in S. Then we can compute the dissimilarity matrix D. If $i \neq j$, $D_{ij} = 1 - S'_{ij}$; otherwise, $D_{ij} = S'_{ij}$, where S' is the normalized symmetric transferring matrix. When $i = j$, the dissimilarity between the same area is 0; when $i \neq j$, the dissimilarity between two areas decreases as the normalized symmetric flow matrix S' increases, which means that high pedestrian flows result in low dissimilarity values, and vice versa.

To verify the effectiveness of iVAT, we apply it to the two synthetic examples. The iVAT image results and reordered contour maps are in Fig. 4. The reordering of these two cases are both Areas {1 9 5 2 7 11 3 10 4 6 8 12}. However, the clustering results are different. For case 1, the clustering is {(1 9 5) (2 7) 11 3 10 4 6 8 12}, i.e., Areas 1, 9 and 5 are strongly related and also 2 and 7. For case 2, the clustering is {(1 9) 5 2 (7 11) (3 10) (4 6) 8 12}. The results indicate that iVAT can cluster the related areas correctly.

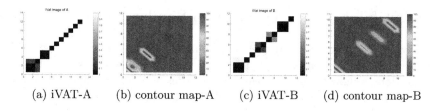

(a) iVAT-A   (b) contour map-A   (c) iVAT-B   (d) contour map-B

Fig. 4. Results on synthetic examples

Table 1. Origin-destination flow matrix

Area	1	2	3	4	5	6	7	8	9	10	11	12	13
13	0	8	0	0	0	0	0	2	0	0	0	0	0
12	1	17	0	0	0	0	0	0	0	0	3	0	0
11	4	102	5	2	0	0	0	11	0	11	0	4	1
10	0	3	1	0	3	0	0	0	0	0	11	1	1
9	0	0	0	0	0	0	0	0	0	0	0	0	0
8	0	16	29	0	6	0	2	0	0	0	14	0	3
7	0	0	0	0	1	0	0	1	0	0	0	0	0
6	0	0	0	0	1	0	0	0	0	0	2	0	0
5	0	7	26	17	0	0	0	4	0	1	0	0	0
4	0	3	8	0	13	0	0	0	0	0	1	0	0
3	0	1	0	8	25	0	0	16	0	2	20	0	0
2	0	0	2	5	5	0	0	32	0	4	103	21	2
1	0	2	0	0	0	0	0	0	0	0	0	0	0

Table 2. Reordered origin-destination matrix

Area	1	11	2	8	3	5	12	4	10	13	6	7	9
9	0	0	0	0	0	0	0	0	0	0	0	0	0
7	0	0	0	1	0	1	0	0	0	0	0	0	0
6	0	2	0	0	0	1	0	0	0	0	0	0	0
13	0	0	8	2	0	0	0	0	0	0	0	0	0
10	0	11	3	0	1	3	1	0	0	1	0	0	0
4	0	1	3	0	8	13	0	0	0	0	0	0	0
12	1	3	17	0	0	0	0	0	0	0	0	0	0
5	0	0	7	4	26	0	0	17	1	0	0	0	0
3	0	20	1	16	0	25	0	8	2	0	0	0	0
8	0	14	16	0	29	6	0	0	0	3	0	2	0
2	0	103	0	32	2	5	21	5	4	2	0	0	0
11	4	0	102	11	5	0	4	2	11	1	0	0	0
1	0	2	0	0	0	0	0	0	0	0	0	0	0

5.2   Case of Real Trajectory Data

Next, we test our method of visual clustering on the real trajectories from one Sunday (20-Jun-2010). We assume that the structure of the scene is known, so we can classify trajectories based on the location of their first (start) and last (end) regions. For example, given $T = \{(x_{start}, y_{start}, t_{start}), ..., (x_{end}, y_{end}, t_{end})\}$ with $(x_{start}, y_{start}, t_{start}) \in A_{11}$ and $(x_{end}, y_{end}, t_{end}) \in A_6$, then T belongs to (11,6).

By counting all the trajectories, we obtain an origin-destination flow matrix (i.e., the frequency-adjacency matrix) F. In this matrix, the value of $F(i, j)$ represents the number of trajectories which start at A_i and end at A_j. Note that the origin-destination matrix can be asymmetric, e.g., Table 1 on Sunday (20-Jun-2010). Applying the iVAT algorithm, there are 12 clusters for all 13 areas, as is shown in Fig. 5. The clustering result is $\{1\ (11\ 2)\ 8\ 3\ 5\ 12\ 4\ 10\ 13\ 6\ 7\ 9\}$.

Using the iVAT results, we obtain the reordered origin-destination flow matrix, which is shown in Table 2. For most origin-destination pairs, there are few trajectories between them (S_{ij} is small compared with S_{max}), leading to $S'_{ij} \sim 0$ and $D_{ij} \sim 1$. Thus, most clusters/area groups contain only one area, except the cluster containing A_{11} and A_2, which means Areas A_2 and A_{11} have high pedestrian flows and they are most related to each other.

6   Identifying Time Periods with Similar Flows

The second question we address in this paper is how to identify which time periods exhibit similar patterns of pedestrian flows. The challenges in addressing this question are how to compare the flow patterns of different time periods, and how to identify which time periods have similar flow patterns. This enables users to profile normal activity.

6.1   Comparing Flow Patterns

Given flow matrices F_i from time period T_i and F_j from time period T_j, we require a measure of how similar are the flows between these two time periods, i.e., we require a distance measure $d(F_i, F_j)$. We use the Frobenius norm which reflects the pairwise difference of individual flows between the same pairs of location, $\|F_i - F_j\|_F = \sqrt{Tr\left[(F_i - F_j)(F_i - F_j)^T\right]}$, where $(F_i - F_j)^T$ is the transpose of $F_i - F_j$, and Tr means the trace of the matrix. For example, given $F_i = \begin{pmatrix} 0 & 5 \\ 3 & 2 \end{pmatrix}$ and $F_j = \begin{pmatrix} 2 & 1 \\ 2 & 4 \end{pmatrix}$, then the Frobenius norm $\|F_i - F_j\|_F = \sqrt{Tr\left[\begin{pmatrix} 20 & -10 \\ -10 & 5 \end{pmatrix}\right]}$.

6.2   Identifying Similar Time Periods

Given a set of flow matrices $F = \{F_1, F_2, ..., F_m\}$ corresponding to m different time periods $T_1, T_2, ..., T_m$, we would like to group or cluster these flow matrices so that we can identify which time periods have similar flow patterns. For example, if F contains seven flow matrices, each corresponding to the average flows on each day of the week, then we would like to detect F in order to identify which days have similar pedestrian traffic.

To achieve this goal, we again make use of the iVAT algorithm. First, we create a $m \times m$ distance matrix D_F, where the $(i, j)^{th}$ entry in D_F is $d(F_i, F_j) = \|F_i - F_j\|_F$. The distance matrix can be normalized by using $norm(D_F) = (D_F - min(D_F))/(max(D_F) - min(D_F))$, where $min(D_F)$ and $max(D_F)$ are the minimum and maximum value in D_F respectively. We then reorder the normalized D_F using iVAT to produce D_F', which should visually reorder the clusters of time periods with similar flow patterns.

We evaluated our proposed method on the data set of 118 days, which has been classified into seven groups, corresponding to 7 days of the week. Then the averaged flow matrices of the 7 days of the week (7 samples, i.e., Sun, Mon, Tue, Wed, Thu, Fri, Sat) are compared. The iVAT results are shown in Fig. 6. The ordering of the iVAT image is {(4 6 3 5 2) (1 7)}, and it shows two clusters, corresponding to clusters of weekdays (Wed, Fri, Tue, Thu, Mon) and the weekend (Sun, Sat).

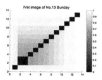

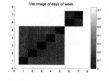

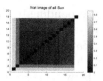

Fig. 5. iVAT of No.13 Sun   **Fig. 6.** iVAT of 7 days   **Fig. 7.** iVAT of all Sun

7   Identifying Anomalous Flow Patterns

Once we have a profile of normal flow patterns over different periods of time, our final question is how to identify which time periods have experienced unusual or anomalous flow patterns. The challenge in addressing this question is how to identify individual time periods in which the pedestrian flows significantly differ from what is expected. This enables users to detect when an anomaly has occurred, and to analyse how the pedestrian flows during that time period differ from what is expected.

Given a set of flow matrices $F = \{F_1, F_2, ..., F_m\}$, the aim of visual anomaly detection is to detect a subset of these flow matrices that are anomalous or outliers compared to the rest. As before, we use the Frobenius norm to compare flow patterns from different time periods, and construct D_F. We then reorder the distance matrix D_F using iVAT to generate D'_F. When we visualize D_F, any anomalous time periods should appear as singleton dark blocks, which are significantly different form the larger clusters in F. For example, consider the set of flow matrices for all Sundays, the iVAT result is shown in Fig. 7. The iVAT result shows that Sunday has four clusters $\{(13)\ (4)\ (8\ 15\ 16\ 5\ 9\ 17\ 18\ 3\ 12\ 11\ 7\ 10\ 1\ 6\ 2)\ (14)\}$, which means that the anomalous time periods are $D' = \{Sun_{13}, Sun_4, Sun_{14}\}$. The average of normal Sundays contour map, and No.4, No.13 and No.14 Sunday contour maps are shown in Fig. 8.

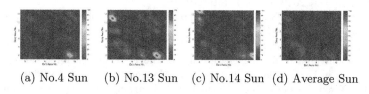

(a) No.4 Sun   (b) No.13 Sun   (c) No.14 Sun   (d) Average Sun

Fig. 8. Contour of abnormal and normal Sundays (Color figure online)

The contour maps indicate that each of the three anomalous Sundays has different high value regions, and all these three time periods are significantly different from the distribution of the average of normal Sundays. For example, in Fig. 8(b), the top left area and lower right area are very bright, which indicates that there are lots of people moving between A_2 and A_{11} as the most visited areas, corresponding to the two largest values $F(11,2) = 102$ and $F(2,11) = 103$ in the origin-destination flow matrix respectively. Some other areas are rather dark, indicating few people moving between these area pairs. There are some regions with relatively bright colour, indicating relatively high value of trajectory numbers between corresponding area pairs, e.g. $A_2 \rightarrow A_8$, $A_5 \rightarrow A_3$, $A_8 \rightarrow A_3$. Also, similar analysis can be applied to other days of week, and the results are omitted to save space.

8   Conclusions and Future Work

We have used the origin-destination matrix to discover and characterize the connectivity between places or regions. In order to find and visualize related areas, we introduce a contour map to represent the origin-destination flow matrix, and propose a visual and parameter-free area clustering method based on the VAT/iVAT algorithms. To detect and visualize abnormal days with significantly different flow patterns, an iVAT based method is also developed. The results on synthetic data and the Edinburgh informatics forum database show that our methods can effectively cluster related areas and identify normal/abnormal pedestrian flow patterns. Possible future research directions are to discuss on scalability of the method on large data and to modify the proposed method for data stream analysis.

Acknowledgments. This work is partially supported by China Scholarship Council. The authors want to acknowledge Mr. Xiaoting Wang for his suggestions on our works and the anonymous reviewers for their constructive suggestions and feedback.

References

1. Bezdek, J.C., Hathaway, R.J.: VAT: a tool for visual assessment of (cluster) tendency. In: IJCNN 2002, pp. 2225–2230 (2002)
2. Chawla, S., Yu, Z., Jiafeng, H.: Inferring the root cause in road traffic anomalies. In: IEEE ICDM 2012, pp. 141–150 (2012)
3. Giannotti, F., Mirco, N., Fabio, P., Dino, P.: Trajectory pattern mining. In: KDD 2007, pp. 330–339 (2007)
4. Liu, L., Assaf, B., Carlo, R.: Urban mobility landscape: real time monitoring of urban mobility patterns. In: Proceedings of the 11th International Conference on Computers in Urban Planning and Urban Management, pp. 1–16 (2009)
5. Liu, W., Xinyi, C., Pengfei, H., Norman, I.B.: Learning motion patterns in unstructured scene based on latent structural information. J. Vis. Lang. Comput. **25**(1), 43–53 (2014)
6. Lu, X., Caixia, W., Nader, K., Arie, C., Anthony, S.: Deriving implicit indoor scene structure with path analysis. In: Proceedings of the 3rd ACM SIGSPATIAL International Workshop on Indoor Spatial Awareness, pp. 43–50 (2011)
7. Majecka, B.: Statistical models of pedestrian behaviour in the forum. Master's thesis, School of Informatics, University of Edinburgh (2009)
8. Pang, L.X., Sanjay, C., Wei, L., Yu, Z.: On detection of emerging anomalous traffic patterns using GPS data. Data Knowl. Eng. **87**, 357–373 (2013)

9. Wang, L., Uyen, T.N., James, C.B., Christopher, A.L., Kotagiri, R.: iVAT and aVAT: enhanced visual analysis for cluster tendency assessment. In: PAKDD 2010, pp. 16–27 (2010)
10. Witayangkurn, A., Teerayut, H., Yoshihide, S., Ryosuke, S.: Anomalous event detection on large-scale GPS data from mobile phones using hidden Markov model and cloud platform. In: UbiComp. 2013, pp. 1219–1228 (2013)

Anomalous Behavior Detection in Crowded Scenes Using Clustering and Spatio-Temporal Features

Meng Yang[1,2(✉)], Sutharshan Rajasegarar[3], Aravinda S. Rao[4],
Christopher Leckie[1,2], and Marimuthu Palaniswami[4]

[1] Department of Computing and Information Systems, The University of
Melbourne, Melbourne, VIC 3010, Australia
myang3@student.unimelb.edu.au,
caleckie@unimelb.edu.au
[2] National ICT Australia (NICTA), Melbourne, VIC 3053, Australia
[3] School of Information Technology, Deakin University,
Melbourne, VIC 3125, Australia
sutharshan.rajasegarar@deakin.edu.au
[4] Department of Electrical and Electronic Engineering, The University
of Melbourne, Melbourne, VIC 3010, Australia
{aravinda.rao,palani}@unimelb.edu.au

Abstract. Anomalous behavior detection in crowded and unanticipated scenarios is an important problem in real-life applications. Detection of anomalous behaviors such as people standing statically and loitering around a place are the focus of this paper. In order to detect anomalous events and objects, ViBe was used for background modeling and object detection at first. Then, a Kalman filter and Hungarian cost algorithm were implemented for tracking and generating trajectories of people. Next, spatio-temporal features were extracted and represented. Finally, hyperspherical clustering was used for anomaly detection in an unsupervised manner. We investigate three different approaches to extracting and representing spatio-temporal features, and we demonstrate the effectiveness of our proposed feature representation on a standard benchmark dataset and a real-life video surveillance environment.

Keywords: Anomaly detection · Spatio-temporal features · Hyperspherical clustering

1 Introduction

Analysis of human behaviour in crowded environment is an important and challenging task for video surveillance. Significant efforts have been made to solve this task, such as using large numbers of surveillance cameras to monitor human behaviour. However, the ubiquity of cameras still causes issues, such as system overload, manual monitoring and low accuracy. Therefore, an automated system for behaviour detection is required to help improve efficiency and reduce detection errors. We aim to detect anomalous events in a target area monitored by cameras over a period of time. Anomalous events

© IFIP International Federation for Information Processing 2016
Published by Springer International Publishing AG 2016. All Rights Reserved
Z. Shi et al. (Eds.): IIP 2016, IFIP AICT 486, pp. 132–141, 2016.
DOI: 10.1007/978-3-319-48390-0_14

include standing statically, loitering around a place, running among a crowd of walking people, and the number of people increasing dramatically at the entrance or exit in some stadium, cinema or other venue. These abnormal events can occur suddenly, hence, an automated and online analysis system is needed for detecting anomalous behaviours.

In this paper, we construct a framework for anomalous behaviour detection, such as remaining static or loitering in the flow of a crowd. This method is almost real-time. In particular, we use a hyperspherical clustering method on the encoded trajectories of pedestrians using novel spatio-temporal feature representations. In other words, after obtaining the tracks of objects and representing the spatial and temporal relationship of these objects, those objects that show behaviour like remaining static or loitering will be detected and declared as anomalous. The main contributions of this work are: (1) our approach to performing object detection and tracking for generating trajectories of objects; (2) we propose the three kinds of spatio-temporal encodings for feature representation, including two novel encoding schemes for anomalous behaviour detection; (3) we use a hyperspherical cluster based distributed anomaly detection method [1] to effectively identify anomalous trajectories in the data; (4) we perform an evaluation on benchmark and real data sets including videos collected from a stadium in Australia; (5) our method is completely unsupervised, hence no labelled data nor supervised training is required.

The datasets used in this work are Melbourne Cricket Ground (MCG) and Performance Evaluation of Tracking and Surveillance (PETS) 2009 videos. The MCG dataset was collected at the Melbourne Cricket Ground with six cameras named C1 to C6. Five of them were installed in a corridor and C1 was placed over a seating area. The total data are 31.05 h video files. PETS2009 is a popular dataset with multi-sensor sequences used for crowd activity recognition [2]. We assume that the video data from cameras are directly available and the cameras are calibrated.

2   Related Work

In order to address the abnormal event detection problem, many algorithms have been proposed in the literature. The methods for anomalous behavior detection can be categorized into two types. One is trajectory analysis and the other is motion representation. Trajectory analysis [3] comprises tracking and distinguishing objects or crowds in the scenes. Motion representation methods analyse patterns such as texture and dynamic models. Optical flow methods are quite popular, for example, Kim and Grauman [4] built a model of optical flow patterns with a mixture of probabilistic Principle Component Analysis (PCA) models, then used a Markov Random Field (MRF) for global consistency guarantees. Mehran [5] learned from crowd behavior studies in [6], and used social force and some other concepts to depict crowd behavior. Then, the concepts and optical flow methods are combined with a latent Dirichlet allocation (LDA) model, and used for anomaly detection. Andrade [7] extracted an optical flow field and used component analysis for reducing dimensionality, and then trained a Hidden Markov Model (HMM) for classifying normal and abnormal behaviors.

Several methods are proposed for anomalous data detection [8]. In particular, supervised and semi-supervised schemes [9] are popular in this category. In terms of unsupervised methods, it is hard to obtain a large amount of pre-labelled data from complicated scenes and predict some unanticipated anomalous data. Recently, in [10], a clustering based anomaly detection method was proposed, which can detect new anomalous data in an unsupervised way.

Since a simple tracking algorithm cannot handle anomalous behaviour detection in crowded scenes, we propose the use of tracking analysis with clustering analysis for anomaly detection. In this work, ViBe [11] based foreground subtraction, Kalman filtering and Hungarian cost algorithm tracking are first used to obtain trajectories of objects. Next, we utilize a fixed-width clustering algorithm, which is an efficient hyperspherical clustering method for abnormal behaviour detection in crowded scenarios. Figure 1 demonstrates our proposed framework for anomalous behaviour detection. The process is divided into six steps: (a) video (camera) inputs, (b) image prepocessing and object detection, (c) object tracking, (d) feature extraction and feature representation, and (e) hyperspherical clustering based anomaly detection. The main challenge we address is how to find a suitable spatio-temporal feature representation in this content.

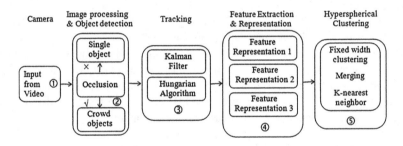

Fig. 1. Overview of our framework for unsupervised anomalous behavior detection

3   The Proposed Approach

Human behaviour has no fixed predefined features, so it is challenging to depict anomalous behaviour directly. Our proposed scheme uses unsupervised clustering to characterize normal behaviour and detect anomalous behaviour clusters. We denote the objects in the scene to be $S = \{s_i : i = 1 \ldots s\}$. For each frame Δ_f, every object s_i has its position in the frame. For several frames (a video sequence), a feature representation consisting of spatio-temporal information is extracted and encoded by three types of schemes, which can depict the trajectory of each object. Further, every object s_i measures a spatio-temporal feature vector $X_i = \{x_k^i : k = 1 \ldots m\}$ and all of the feature vectors build the spatio-temporal feature matrix $X = \cup_{i=1\ldots n} X_i$. Finally, X is the input to a hyperspherical based clustering algothrithm [1] for outlier detection, and the detected outliers are regarded as the anomalous behaviours.

In this work, at first, video preprocessing is performed followed by object detection, which is based on ViBe background subtraction. Next, the objects are tracked by a Kalman filter and Hungarian algorithm. We then propose three alternative methods for feature representation so that clustering can then be applied to the trajectories. Finally, feature vectors are categorized into normal and anomalous clusters by using fixed width clustering. Overall performance of the framework depends on the object detection, tracking, feature representation and clustering methods.

3.1 Image Preprocessing and Object Detection

The video frames are converted to grayscale images at first. Then, these grayscale images are filtered by a 2D Gaussian low-pass filter in order to filter the high-frequency noise. The parameter of the Gaussian low-pass filter is set to be $\sigma = 0.5$ and the block size is 5×5. The choice of filter parameters is based on maintaining a part of the edge information while at the same time keeping low-frequency information.

The object detection algorithms can be categorized into three groups: (1) frame difference method, (2) optical flow based approach, (3) background subtraction. Especially, background subtraction is a crucial category of object detection method and several techniques have been proposed in the literature. These methods can be divided into parametric methods and sample-based methods. The former methods are based on the location of each pixel and the latter one is based on aggregating previously observed values for each pixel location. Considering the complexity of the scenarios and the sensitivity of the environment, in this work, we choose a sample-based algorithm ViBe [11] for background modeling and subtraction. The algorithm begins with defining a pixel model by using a set of sample values that can be used for background estimation, then updating the model regularly. The objects can be detected by subtracting the pixel model from incoming frames.

3.2 Tracking

Kalman filtering [12] is used for tracking detected objects by the previous step. The process of Kalman filtering is: Initialization-> Prediction-> Correction. The motion equation and model are displayed below:

$$x_{k+1} = F_{k+1,k}x_k + w_k \tag{1}$$

$$\begin{bmatrix} x_{k+1} \\ y_{k+1} \\ \Delta x_{k+1} \\ \Delta y_{k+1} \end{bmatrix} = \begin{bmatrix} 1 & 0 & 1 & 0 \\ 0 & 1 & 0 & 1 \\ 0 & 0 & 1 & 0 \\ 0 & 0 & 0 & 1 \end{bmatrix} \begin{bmatrix} x_k \\ y_k \\ \Delta x_k \\ \Delta y_k \end{bmatrix} + w_k \tag{2}$$

and the measurement equation and model are given below

$$y_k = H_k x_k + v_k \tag{3}$$

$$\begin{bmatrix} xm_k \\ ym_k \end{bmatrix} = \begin{bmatrix} 1 & 0 & 0 & 0 \\ 0 & 1 & 0 & 0 \end{bmatrix} \begin{bmatrix} x_k \\ y_k \\ \Delta x_k \\ \Delta y_k \end{bmatrix} + v_k \qquad (4)$$

where, $F_{k+1,k}$ is the transition matrix and w_k is additive process noise; H_k is the measurement matrix and v_k is the measurement noise; and xm_k are ym_k the observed position. Because of the crowded environment, the tracking algorithm we use should handle multi-object association and assignment. Therefore, we use the Hungarian algorithm [13] for multiple objects tracking.

3.3   Feature Extraction and Representation

For the video analytics, anomalous crowds can be detected at three levels: temporal, spatial and spatio-temporal, which is a combination of the spatial and temporal levels. Crowded scenes contain a large number of objects with different behaviours within one frame, and these behaviours occur and change within the image sequence. In this case, anomalous crowd behaviour should be analysed on both the spatial and temporal levels. Therefore, it would be better to choose a spatio-temporal feature representation as our encoding scheme in this scenario.

After the feature extraction and representation step, an appropriate feature representation is needed for the spatio-temporal trajectories, which can have arbitrary length. This feature representation becomes the input of clustering in order to identify normal and abnormal behaviour. In this work, we use three types of feature representation schemes for crowd anomaly detection. The three schemes are all spatio-temporal representations, including one representation which is modified from the coding scheme in [14], and two other novel feature representations. The details of the feature and representation methods are given below.

Let m be the width and n be the height of the video frame. The input frame is divided into bx × by blocks, where $1 < bx \leq m$, $1 < by \leq n$. Block size selection affects the encoding results, which will be discussed in the next section. Next, a feature representation with spatio-temporal information is extracted for each object. The block value starts from 1 for the original position of each object. In the following frames, the object will either stay in the same block or move to another block. The following three encoding schemes are used to represent the status of each object.

(1) Feature Representation 1 (FR1): If the object enters into another block, then this new block value will be increased by one. If the object stays in the same block, the value of that block will not be changed. Further, if this object enters into the same block again after several steps, the new value will replace the old one.

(2) Feature Representation 2 (FR2): The value of each block depends on the length of time the object stays in that block. If the object stays in one block for 10 frames and moves to another block for 1 frame, then the feature value of these two blocks are 10 and 1, respectively.

(3) Feature Representation 3 (FR3): This is a combination of the former two types. At first, an object enters into a block and this block will be assigned the value one,

then if this object stays in the same block, the value will be the number of frames that the object stays in that block. Further, if the object moves to another block, the value of this new block will start from the value of the former block plus one.

Once we have encoded the frames and the blocks as mentioned above, we will have a collection of trajectories for the objects in the video over the period of time considered for analysis. The next step is to identify trajectories that are normal or anomalous. Next we describe a hypespherical clustering scheme to perform this in an unsupervised way.

3.4   Clustering Based Anomaly Detection

In this work, the clustering algorithm we used is from [1]. The method has three main steps: (1) Fixed width clustering, which is a hyperspherical clustering with fixed radius. The process is as follows. The first cluster is created with a fixed width (radius), centred at the first data point. Then the distance between the next new data vector and its closest cluster centre is calculated. If the distance is less than the radius, then the data vector is added to this cluster and the centroid of that cluster is recalculated as the mean of all data vectors in it. If the distance is more than the radius, then the data vector will form a new cluster using it as the centroid. The process is continued until all the data vectors are considered. (2) Merging, When the distance between the center of two clusters is less than a threshold τ, then they will be merged into one new cluster. (3) The anomalous clusters are identified by using the K nearest neighbor (K-NN) approach [15]. This method helps cluster similar behaviors (trajectories) and finds the anomalous events in the video in an unsupervised manner. The parameters of this algorithm are ω (cluster width) and ψ (the number of standard deviations of the inter-cluster distances used for identifying the anomalous clusters). The number of clusters yielded is based on ω, and ψ determines the sensitivity of anomaly detection. The detailed steps of this method and the parameters can be found in [1].

4   Results and Discussion

In this section, we discuss the dataset used, and then show the anomaly detection results based on the three types of encoding schemes proposed. The details of the data we used are listed in Table 1. In the data we used for evaluation, people who stand statically or loiter in the scene for a while are regarded as anomalies. We manually annotated the anomalous objects as the ground truth. The computer vision part is implemented using OpenCV 3.0 with Visual Studio 2013, and the anomaly detection part is implemented in Java. The computer used is Windows 7 (64 bit) consisting of an Intel® i7 - 4790 CPU running at 3.6 GHz with 16 GB RAM. The computer also includes a 4 GB NVIDIA NVS 315 HD 4600 graphics card.

Table 1. Detailes of the data used, including MCG and PETS2009 dataset.

MCG Dataset			PETS2009 Dataset			
Date	Camera	TL	Camera	TS	View	Frames
23-Sep-2011	C2–C6	22:01	S2–L1	12–34	001	794
24-Sep-2011	C2–C6	14:01	S2–L2	14–55	001	435
24-Sep-2011	C6_cutl–12	00:20	S2–L3	14–41	001	239

※ TL = TimeLength (mm:ss) ※ TS = Timestamp

4.1 Object Detection Performance

Loitering and static objects are detected based on using the three types of spatio-temporal feature representations (FR1, FR2, FR3). Loitering objects are those who walk around and come back to the same place. Static objects are those who stand statically or move slightly for a while in the scene. The number of identified objects was compared to the ground truth generated by annotating the original video manually. For these three feature representation schemes, they have similar feature matrices. The maximum value of the row vector increased with the occurrence of anomalous objects. The results are tabulated in Table 2 (MCG dataset) and Table 3 (PETS2009 dataset) respectively. For the hyperspherical clustering part, the parameters were set as: $\omega = 5$, $\tau = \frac{1}{2}\omega$, K-NN = 3 and $\psi = 1$.

Table 2. The table lists the number of detected anomalous objects using the three types of spatio-temporal feature encoding schemes on the MCG dataset (24-Sep-2011). Left-top section is for C2, right-top section is for C3, left-bottom is for C5 and right-bottom is for C6.

GT = 206	8 × 8	16 × 16	32 × 32	GT = 93	8 × 8	16 × 16	32 × 32
FR1	84	150	108	FR1	23	44	58
FR2	193	196	209	FR2	74	74	81
FR3	167	227	181	FR3	62	73	65
GT = 118	8 × 8	16 × 16	32 × 32	GT = 66	8 × 8	16 × 16	32 × 32
FR1	48	76	96	FR1	16	50	63
FR2	96	140	125	FR2	54	61	56
FR3	98	128	106	FR3	51	58	58

※ GT = Ground Truth

Table 3. The table lists the number of detected anomalous objects using the three types of spatio-temporal features encoding schemes on the PETS2009 dataset.

GT:8	8 × 8	16 × 16	32 × 32	GT:6	8 × 8	16 × 16	32 × 32	GT:3	8 × 8	16 × 16	32 × 32
FR1	6	4	2	FR1	5	5	3	FR1	2	2	2
FR2	9	5	3	FR2	8	4	7	FR2	2	2	1
FR3	7	6	2	FR3	6	5	4	FR3	3	4	2

For the MCG dataset, the anomalies detected by the three types of spatio-temporal feature representation have been tabulated in Table 2. Videos from camera C2, C3, C5 and C6 (24-September-2011) were used, which are all of length 14 min and 1 s. These frames were divided into 8×8, 16×16 and 32×32 blocks. The ground truth was generated by annotating the video files. From Table 2, we find that the number of anomalies is affected by the block size selection. For example, comparing the 8×8 and 16×16 blocks, the number of anomalies detected by the former one is larger than the latter one. This means that the larger block size results in lower resolution coding, so the number of detected anomalous objects are less. However, this is not correct for all cases.

Further, the number of detected anomalies based on FR1 is less than for FR2 and FR3. This can be explained by using an example. If there is an object standing statically in block 2 for 100 frames and then walks to block 3, then the value of block 2 should be 1 and block 3 should be 2 by using FR1. Next, the value of block 2 should be 100 and block 3 should be 1 based on FR2, and the two blocks will be 100 and 101 based on FR3. It is clear that the object will be regarded as abnormal byusing FR2 and FR3, whereas normal in FR1. The large number of static objects in the MCG dataset causes the lower number of detected anomalies based on FR1. In other words, FR2 and FR3 are more suitable for detecting static objects.

For the PETS2009 dataset, from Table 3, it can be seen that the number of anomalies detected based on FR1 is similar with FR2 and FR3 in Table 3. This is because the anomalous behavior in the PETS2009 dataset is different from the MCG dataset. The anomalous objects loiter on a large scale in PEST2009, whereas objects stand statically or move only slightly in the MCG. We can assume that the three types of feature representation methods have similar effects on detecting loitering objects.

From Tables 2 and 3, we find that for some instances the number of detected anomalies is more than the ground truth. There are two possibilities that can cause this situation. One is that normal objects are regarded as anomalies, and the other is because of the algorithm we used for extracting trajectories. The object detection and tracking algorithms we used are not based on a pre-trained model, which means we cannot always obtain a complete and smooth trajectory of one object. The trajectory of the object can be split into 2 or 3 parts, which causes the situation that one object is detected as 2 or 3 abnormal objects.

4.2   Accuracy Analysis

The MCG dataset was used for our accuracy evaluation, including C2-C6 (23-Sep-2011), C2-C6 (24-Sep-2011) and 12 cuts of video of C6 (24-Sep-2011). In the last section, we obtained the number of detected anomalous objects, while we also obtained the object ID, which can denote the real behavior of his/her (normal or abnormal) movement pattern. We choose 8×8 as the block size for this evaluation part. Finally, we can obtain the detection accuracy of the detected anomalous objects. The results have been tabulated in Table 4.

From Table 4(a), it is clear that the accuracy of FR1 is lower than FR2 and FR3, which is similar to the result of Table 2. FR2 and FR3 have similar accuracy. From

Table 4. The detection accuracy using the three types of spatio-temporal features encoding schemes. (a) C2-C6 (23-Sep-2011) and C2-C6 (24-Sep-2011), (b) 12 video cuts of C6 (24-Sep-2011)

Dataset	23-Sep-2011				24-Sep-2011				SD
Accuracy	C2	C3	C5	C6	C2	C3	C5	C6	
FR1	40.3%	31.4%	39.0%	22.7%	48.5%	41.6%	44.6%	44.8%	39.1 ± 7.8%
FR2	78.6%	75.3%	76.3%	74.2%	75.8%	74.2%	77.7%	79.2%	76.4 ± 1.8%
FR3	72.3%	65.6%	78.0%	71.2%	74.3%	69.9%	76.2%	78.4%	73.2 ± 4.1%

(a)

Dataset	C1	C2	C3	C4	C5	C6
FR1	66.7%	60%	57.1%	50%	44.4%	80%
FR2	83.3%	100%	71.4%	87.5%	77.8%	80%
FR3	100%	80%	71.4%	87.5%	88.9%	80%

Dataset	C7	C8	C9	C10	C11	C12	SD
FR1	75%	66.7%	40%	75%	60%	66.7%	63.5 ± 9.9%
FR2	87.5%	77.8%	80%	100%	80%	100%	85.4 ± 9.4%
FR3	87.5%	77.8%	60%	100%	100%	100%	86.1 ± 12.3%

※SD = Standard deviation

(b)

Table 4(b), although the accuracy of FR1 is still lower than FR2 and FR3, the accuracy of FR1 is increased. This is because the time length of the video cut is short, so the time length of loitering is short. The feature matrix of these three feature representation methods has no major differences. In terms of the false positive rate, it is around 10 % to 20 % for FR1, FR2 and FR3. The effect is quite similar among the three encoding schemes. Generally speaking, FR2 and FR3 are more suitable for detecting static objects, especially in a crowded scene. The three types of feature representation have similar effects on detecting loitering objects.

5   Conclusion

Anomalous behavior detection in crowded and unanticipated scenarios is an important problem in real-life applications. In this work, the anomalous behaviors of standing statically and loitering in a video were detected by using two novel encoding schemes for spatio-temporal features. At first, ViBe was used for object detection. Then, Kalman filtering and a Hungarian cost algorithm were implemented for multi-object tracking. Next, the spatio-temporal features were extracted and represented by three types of schemes. In the end, a hyperspherical clustering based algorithm was used for anomaly detection. The evaluation reveals that our proposed unsupervised anomaly detection scheme using our novel spatio-temporal features is capable of detecting anomalous events such as loitering and stationary objects with high accuracy on a real life and a benchmark dataset.

Acknowledgements. This work was supported by National ICT Australia (NICTA).

References

1. Rajasegarar, S., Leckie, C., Palaniswami, M.: Hyperspherical cluster based distributed anomaly detection in wireless sensor networks. J. Parallel Distrib. Comput. **74**, 1833–1847 (2014)
2. Ferryman, J., Crowley, J., Shahrokni, A.: PETS 2009 Benchmark data (2009). http://www.cvg.rdg.ac.uk/PETS2009/a.html
3. Stauffer, C., Grimson, W.E.L.: Learning patterns of activity using real-time tracking. IEEE Trans. Pattern Anal. Mach. Intell. **22**, 747–757 (2000)
4. Kim, J., Grauman, K.: Observe locally, infer globally: a space-time MRF for detecting abnormal activities with incremental updates. In: IEEE Conference on Computer Vision and Pattern Recognition, (CVPR 2009), pp. 2921–2928 (2009)
5. Mehran, R., Oyama, A., Shah, M.: Abnormal crowd behavior detection using social force model. In: IEEE Conference on Computer Vision and Pattern Recognition, (CVPR 2009), pp. 935–942 (2009)
6. Helbing, D., Molnar, P.: Social force model for pedestrian dynamics. Phys. Rev. E **51**, 4282 (1995)
7. Andrade, E.L., Blunsden, S., Fisher, R.B.: Modelling crowd scenes for event detection. In: 18th International Conference on Pattern Recognition, (ICPR 2006), pp. 175–178 (2006)
8. Rajasegarar, S., Leckie, C., Palaniswami, M.: Anomaly detection in wireless sensor networks. IEEE Wirel. Commun. **15**, 34–40 (2008)
9. Chandola, V., Banerjee, A., Kumar, V.: Anomaly detection: a survey. ACM Comput. Surv. (CSUR) **41**, 15 (2009)
10. Rajasegarar, S., Leckie, C., Palaniswami, M., Bezdek, J.C.: Quarter sphere based distributed anomaly detection in wireless sensor networks. In: IEEE ICC, pp. 3864–3869 (2007)
11. Barnich, O., Van Droogenbroeck, M.: ViBe: a powerful random technique to estimate the background in video sequences. In: IEEE International Conference on Acoustics, Speech and Signal Processing, (ICASSP 2009), pp. 945–948 (2009)
12. Cuevas, E.V., Zaldivar, D., Rojas, R.: Kalman filter for vision tracking (2005)
13. Kuhn, H.W.: The Hungarian method for the assignment problem. Naval Res. Logistics Q. **2**, 83–97 (1955)
14. Rao, A.S., Gubbi, J., Rajasegarar, S., Marusic, S., Palaniswami, M.: Detection of anomalous crowd behaviour using hyperspherical clustering. In: International Conference on, Digital Image Computing: Techniques and Applications (DICTA), pp. 1–8 (2014)
15. Ramaswamy, S., Rastogi, R., Shim, K.: Efficient algorithms for mining outliers from large data sets. In: ACM SIGMOD Record, pp. 427–438 (2000)

An Improved Genetic-Based Link Clustering for Overlapping Community Detection

Yong Zhou and Guibin Sun[✉]

School of Computer Science and Technology, China University of Mining and Technology,
Xuzhou 221008, China
yzhou@cumt.edu.cn, sunguibinbest@qq.com

Abstract. The problem of community detection in complex networks has been intensively investigated in recent years. And it was found that the communities of complex networks often overlap with each other. So in this paper, we propose an improved genetic-based link clustering for overlapping community detection. The first, the algorithm changes the node graph into the link graph. The second, the algorithm adopts the genetic algorithm to detect the link communities. The Third, the algorithm transforms the link communities into the node communities. Automatically, the nodes, which are linked with edges belonged to different link communities, will be the overlapping nodes. The last, in order to improve the quality of community detection, we define an effective method to solve the "excessive overlap" problem. The experimental results shows that the proposed algorithm is effective and efficient on both simulate networks and real networks.

Keywords: Genetic-based · Link clustering · Overlapping communities · Community detection

1 Introduction

Many complex systems in nature and society can be described in terms of networks or graphs. The study of networks is crucial to understanding both the structure and the function of these complex systems. Researchers found that a common feature which is called community structure exists in many complex networks. Community structures are always expressed as clusters of nodes with dense connections within cluster and sparse connections with the other clusters. The community structure plays an important role in the complex network which can help people to understand the function of the complex network and find the potential law in the complex network. Take the World Wide Web as an example, close hyperlink web pages form a community and they often talk about related topics.

The identification of community structure has attracted much attention from various scientific fields. A lot of algorithms have been proposed for detecting communities in complex networks. The traditional community detection algorithm is to divide the complex network into several disconnected communities (or clusters,

© IFIP International Federation for Information Processing 2016
Published by Springer International Publishing AG 2016. All Rights Reserved
Z. Shi et al. (Eds.): IIP 2016, IFIP AICT 486, pp. 142–151, 2016.
DOI: 10.1007/978-3-319-48390-0_15

groups, etc.), and each node must be affiliated with one community. The representative algorithms include the modularity optimization algorithm [1, 2], spectral clustering method [3, 4], and so on. However, there are many overlapping networks in real world. That is to say, in the complex networks, some nodes can't belong to only one community, they can belong to multiple communities at the same time. For example, in a social network, each person can belong to more than one social group at the same time (e.g., school, family, friends, etc.).

Recently, the overlapping community structure has been widely studied. Some algorithms use the clique percolation to detect the overlapping community, such as the well-known CPM [5], SCP [6] and EAGLE [7]. Some algorithms utilize the local expansion by optimizing a local benefit function, such as LFM [8], MONC [9], CIS [10] and OSLOM [11]. Some label propagation based algorithms allow multiple labels for each node to detect overlapping structure, such as COPRA [12], SLPA [13], etc. Some algorithms are Based on the link clustering, such as LINK [14], Link Maximum Likelihood [15] and Link-Comm [16]. Although the overlapping community detection has obtained significant achievements, with the network structure increasingly complex, the community detection is more difficult. how to more accurately and effectively detect the overlapping community structure is still a great challenge.

In this paper, we propose an improved genetic-based link clustering for overlapping community detection. Firstly, the algorithm changed the node graph into the link graph. Secondly, the algorithm adopted the genetic algorithm to detect the link communities. Thirdly, the algorithm transformed the link communities into the node communities. Automatically, the nodes, which are linked with edges belonged to different link communities, will be the overlapping nodes. Last, in order to improve the quality of community detection, we defined an effective method to solve the "excessive overlap" problem. The effectiveness of the proposed algorithm is demonstrated by extensive tests on both simulate networks and real networks with a known community structure. Through experimental comparison, the proposed algorithm is effective and efficient in overlapping community detection.

2 Related Work

The genetic algorithm for overlapping community detection (GaoCD) [17] was newly proposed in 2013. In this paper, they proposed a genetic algorithm for overlapping community detection based on the link clustering. Different from those node-based overlapping community detection algorithms, the GaoCD algorithm applies a novel genetic algorithm to cluster on the edge set of network. The genetic representation and the corresponding operators effectively represent the link communities and make the number of the communities determined automatically. In the GaoCD algorithm, it mainly includes three components: objective function, genetic representation and genetic operators.

2.1   Objective Function

In the GaoCD algorithm, the partition density D is utilized to evaluate the link density within communities. The partition density D is proposed in the LINK algorithm [10], which emphasizes the community density and ignores the connection among communities. the partition density D is defined as follows.

For a network with M links and N nodes, $P = \{P_1, P_2,..., P_c\}$ is a partition of the links into C subsets. The number of links in subset P_c is m_c. The number of induced nodes, all nodes that those links touch, is n_c.

$$D = \frac{2}{M} \sum_c m_c \frac{m_c - (n_c - 1)}{(n_c - 2)(n_c - 1)} \tag{1}$$

2.2   Objective Function

In the GaoCD algorithm, a gene represents a link. An individual gene sequence in the population is represented as a gene type $[g_0, g_1,..., g_i, ..., g_{m-1}]$. Among them, the m is the number of the edges in the network, $i \in [0, m)$ is the identifier of edges in the network, and each g_i is a random adjacent edge of edge i.

For example, in Fig. 1(a), e_0 has two adjacent edges e_1 and e_2. So the e_1 is the possible value of g_0. The encoding schema guarantees that every community partition can be encoded into a corresponding gene type and every gene type can be decode into an valid community partition. What' s more, the encoding schema can automatically determine the number of the communities, without any prior information.

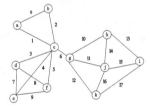

Index	0	1	2	3	4	5	6	7	8	9	10	11	12	13	14	15	16	17
GeneType	2	6	1	7	5	8	11	9	7	5	14	12	17	15	16	17	12	13

(a) The example network   (b) The genetic representation

Fig. 1.   Illustration of the genetic representation

2.3   Genetic Operation

According to the genetic representation, The GaoCD algorithm adopts the corresponding genetic operators.

In the crossover operation, They randomly select two individuals from the current population. The exchanging positions are randomly generated and then exchange the genes in these positions between these two individuals. Since the g_i is always the identity of the adjacent edges of e_i, the exchanged individuals also follow the genetic representation rule: g_i is an adjacent edge of e_i.

In the mutation operation, an individual is randomly selected from the current population and the positions are randomly generated. Then they reassign the gene values on these positions with a random adjacent edge.

3   An Improved Genetic-Based Link Clustering for Overlapping Community Detection

The GaoCD algorithm can effectively reveal overlapping structure. However, the GaoCD algorithm is also easy to appear the "excessive overlap" problem.

For example, the Fig. 2. is two kinds schematic diagrams of the "excessive overlap" problem. In the Fig. 2(a), all nodes should be divided into only one community, however, they are divided into two communities, making the node e and node b become the overlapping nodes. In the Fig. 2(b), the node e should only belong to the right community, however, it belongs to the both right community and left community.

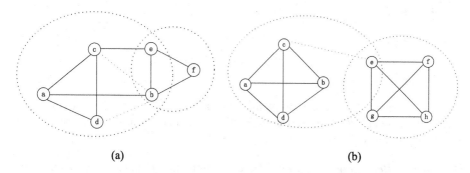

(a)   (b)

Fig. 2. Illustration of the "excessive overlap" problem

In order to avoid the "excessive overlap" problem and improve the community detection performance. we proposed an improved genetic-based link clustering for overlapping community detection.

3.1   Community Similarity

To solve the "excessive overlap" problem as shown in Fig. 2(a), we define a community similarity to measure the contact ratio of communities.

Definition 1 *Community Similarity.* Given two communities $C1$ and $C2$, the community similarity is define as

$$S(C_1, C_2) = \frac{|C_1 \cap C_2|}{\min(|C_1|, |C_2|)} \tag{2}$$

Given a set of communities *CS* and a community *C*, we can define the near-duplicates of *C* to be all communities in *CS* that are within a contact ratio Δ, where Δ is the maximum community similarity threshold. When the community similarity of the two communities is beyond threshold Δ, the two communities will be merged. In the experiment, we found that it is very reasonable that this threshold Δ be set at about 0.66. The algorithm of calculating community similarity is shown in algorithm 1.

```
Algorithm1. CommSim
01 Input: a set of communities CS, the threshold Δ.
02 Output: updated CS.
03 for each i from 0 to CS.size() do
04   for each j from i +1 to CS.size() do
05     temp = S(Cᵢ, Cⱼ); // use the formula(2).
06     if temp > Δ
07       Cᵢ = merge(Cᵢ, Cⱼ);
08       delete Cⱼ from CS;
09     end
10   end
11 end
12 return CS;
```

3.2 Belonging Coefficients

To solve the "excessive overlap" problem as shown in Fig. 2 (b), we define the belonging coefficients to decide that the overlapping nodes belong to multiple communities or only a single community.

Definition 2 *Belonging Coefficients*. Given a community *C* and an overlapping node *v*, belonging coefficients is defined as fallow:

$$BC(v, C) = \frac{|E(v)||E(v)|}{|E(C)||K(v)|} \tag{3}$$

Among them, the node v denotes an overlapping node which belongs to community C. $E(v)$ denotes the edges which connect node v to the community C. $E(C)$ denotes the edges in the community C. $K(v)$ denotes the degree of node v.

For the nodes with multiple memberships, we use the belonging coefficients to determine whether nodes are excessive overlap nodes. In order to facilitate comparison, we introduce a threshold τ, where τ is the maximum difference between two belonging coefficients in different communities. In the experiment, we found that it is very reasonable that this threshold τ be set at about 0.25. The algorithm about the belonging coefficients is shown in algorithm 2.

```
Algorithm2. BelongCoefficient
01 Input: communities CS, overlapping nodes NS,
   threshold τ.
02 Output: updated CS.
03 for each i from 0 to NS.size() do
04    get the communities C related to the node i;
05    T = ∅; //The set T is used to store the belonging
         coefficients related to the node i.
06    for each j from 0 to C.size() do
07        Tj = BC(i, Cj); // use the formula(4).
08    end
09    sort(T); //sorting the set T and corresponding
         communities C.
10    max = T0; //maximum belonging coefficient.
11    for each k from 1 to T.size() do
12            if T0 - Tk > τ
13                delete the overlapping node i from relating
         community Ck, Ck+1, ..., CT.size()-1;
14                break;
15            end
16    end
17 end
18 return CS;
```

4 Experiments

In this section, the IGLC algorithm is tested on the simulated data sets and real data sets, respectively. Experimental environment: Processor Inter (R) Core (TM) i5 3.1 GHz PC, memory 4G, the operating system is Windows 7, programming environment Matlab R2009a.

4.1 Experimental Data Sets

The Simulated Data Sets. Currently, the LFR benchmark network [18, 19] is the most commonly used data set in community detection. We generate two LFR benchmark networks, whose detail information are shown in Table 1. Some important parameters of the benchmark networks are as follow:

Table 1. The LFR benchmark networks

Num	N	k	maxk	minc	maxc	on	mu	om
S1	1000	20	50	10	50	100	0.1	2 ~ 8
S2	1000	20	50	10	50	100	0.3	2 ~ 8

N: the number of nodes; k: the average degree; *maxk*: the maximum degree; *minc*: the minimum for the community sizes; *maxc*: the maximum for the community sizes; *on*: the number of overlapping nodes; *mu*: mixing degree; *om*: the number of communities that each node can belong to;

The Real Data Sets. We make experiments on five well known social networks, whose real community structure have been given. Their specific information is shown in Table 2.

Table 2. The real network

Name	Nodes	Edges	Source
Karate	34	78	[20]
Dolphins	62	159	[20]
Political Books	105	441	[20]
Football	115	613	[20]
Netscience	379	914	[20]

4.2   Evaluation Criteria

In the experiments, we use the evaluation criteria normalized mutual information (*NMI*) [12] and extended modularity (*EQ*) [7] to evaluate the communities.

Normalized Mutual Information (NMI). The *NMI* is used to measure similarity between the results of algorithm with true class values.

Assuming that the true class values of the data sets are $C = \{C_1, C_2, ..., C_k\}$, and the class labels obtained by the algorithm are $U = \{U_1, U_2, ..., U_l\}$, where k and l denote the number of clusters in C and U. The number of nodes in the Ci $(1 \leq i \leq k)$ and Uj $(1 \leq j \leq l)$ are n_i and n_j respectively. The length of intersection of C_i and U_j is n_{ij}, so *NMI* is defined as Eqs. (4-5).

$$NMI = \frac{2 \times I(C, U)}{H(C) + H(U)} \tag{4}$$

$$NMI = \frac{-2 \sum_{i=1}^{k} \sum_{j=1}^{l} \frac{n_{ij}}{n} \log \frac{n_{ij}}{n_i \times n_j}}{\sum_{i=1}^{k} n_i \log \frac{n_i}{n} + \sum_{j=1}^{l} n_j \log \frac{n_j}{n}} \tag{5}$$

Extended Modularity (EQ). The *EQ* is a variant of the commonly used modularity (*Q*) metric [1], which is defined for overlapping communities by Shen. This extended modularity is defined as follow:

$$EQ = \frac{1}{2m} \sum_{C} \sum_{i,j \in C_k} \frac{1}{O_i O_j} [A_{ij} - \frac{k_i k_j}{2m}] \tag{6}$$

4.3 Experimental Results

In the experiments, we use two algorithms to compare with the proposed IGLC algorithm. The two algorithms are COPR [12] and LINK [14], respectively. The parameters of IGLC are set as follows: $size = 100$, $gens = 100$, $pc = 0.6$, $pm = 0.4$, $\Delta = 0.66$, $\tau = 0.25$. The parameter of COPRA is set as follows: $v = 4$. The LINK algorithm don't need parameters.

The Results on the Simulated Data Sets. The results on the two simulated data sets are shown in the Fig. 3. The abscissa is the *om* whose value ranges from 2 to 8, and the ordinate is the *NMI* value. The NMI value of per *om* is the average of 10 times.

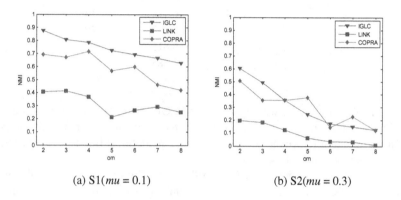

(a) S1(*mu* = 0.1)   (b) S2(*mu* = 0.3)

Fig. 3. The results on the simulated data sets

(1) *Compared with the LINK algorithm.*
    In the two LFR benchmark networks, the results of IGLC are all better than the results of LINK. Because the LINK algorithm exists "excessive overlap" problem, which seriously reduces the quality of community detection. However, the proposed IGLC algorithm solves the "excessive overlap" problem very well. Therefore, the proposed IGLC algorithm effectively improves the quality of community detection.
(2) *Compared with the COPRA algorithm.*
    In the low mixing degree LFR benchmark network(*mu* = 0.1), the *NMI* values of IGLC are all better than the *NMI* values of COPRA. In the high mixed degree LFR benchmark network (*mu* = 0.3), the *NMI* values of IGLC are most better than the *NMI* values of COPRA, Only in a few cases, the COPRA has higher NMI value (e.g. *om* = (5, 7) in the S1. In addition, with the increase of *om*, the community detection is becoming more and more difficult. The *NMI* value of COPRA present fluctuations, which demonstrates that the COPRA algorithm has poor robustness. However, the *NMI* value of proposed IGLC algorithm present the steady downward trend, which demonstrates that our algorithm has good robustness.

In conclusion, the proposed IGLC algorithm can obtain better quality of overlapping community detection compared with the LINK algorithm and COPRA algorithms in most cases.

The Results on the Real Data Sets. Table 3 shows community detection results of three algorithms on real data sets, whose detailed information is shown in Table 2. The bold in each row is the optimal community detection result. The evaluation criterion in Table 3 is the extended modularity EQ.

Table 3. The results on the real network

Name	LINK	COPRA	IGLC
Karate	0.146	0.423	**0.514**
Dolphins	0.351	0.683	**0.729**
Political Books	0.254	**0.813**	0.804
Football	0.557	0.685	**0.689**
Netscience	0.457	0.812	**0.893**

In the Table 3, compared with the LINK and COPRA algorithm, the proposed IGLC algorithm can obtain better clustering results on most networks. Although the proposed IGLC algorithm don't have the best EQ values on the Political Books network, The EQ values are also the second best values.

In conclusion, the proposed IGLC algorithm can achieve acceptable results on real data sets, so the IGLC algorithm is reasonable and effective in overlapping community detection.

5   Conclusion

In this paper, we propose an improved genetic-based link clustering for overlapping community detection(IGLC). The IGLC algorithm mainly includes two parts. One part is adopting the GaoCD algorithm to detect the link communities. The other is transforming the link communities into the node communities and adopting the community similarity and the belonging coefficients to solve the "excessive overlap" problem. Through experimental comparison, the proposed algorithm is effective and efficient on both simulate networks and real networks.

References

1. Newman, M.E.J., Girvan, M.: Finding and evaluating community structure in networks. Phys. Rev. E **69**(2), 026113 (2004)
2. Lee, J., Gross, S.P., Lee, J.: Modularity optimization by conformational space annealing. Phys. Rev. E **85**(5), 056702 (2012)
3. Shen, H.W., Cheng, X.Q.: Spectral methods for the detection of network community structure: a comparative analysis. J. Stat. Mech: Theory Exp. **10**, P10020 (2010)

4. Jiang, J.Q., Dress, A.W.M., Yang, G.: A spectral clustering-based framework for detecting community structures in complex networks. Appl. Math. Lett. **22**(9), 1479–1482 (2009)
5. Palla, G., Derényi, I., Farkas, I., et al.: Uncovering the overlapping community structure of complex networks in nature and society. Nature **435**(7043), 814–818 (2005)
6. Kumpula, J.M., Kivelä, M., Kaski, K., et al.: Sequential algorithm for fast clique percolation. Phys. Rev. E **78**(2), 026109 (2008)   -
7. Shen, H., Cheng, X., Cai, K., et al.: Detect overlapping and hierarchical community structure in networks. Physica A **388**(8), 1706–1712 (2009)
8. Lancichinetti, A., Fortunato, S., Kertész, J.: Detecting the overlapping and hierarchical community structure in complex networks. New J. Phys. **11**(3), 033015 (2009)
9. Havemann, F., Heinz, M., Struck, A., et al.: Identification of overlapping communities and their hierarchy by locally calculating community-changing resolution levels. J. Stat. Mech: Theory Exp. **2011**(01), P01023 (2011)
10. Kelley S.: The existence and discovery of overlapping communities in large-scale networks. Rensselaer Polytechnic Institute (2009)
11. Lancichinetti, A., Radicchi, F., Ramasco, J.J., et al.: Finding statistically significant communities in networks. PLoS ONE **6**(4), e18961 (2011)
12. Gregory, S.: Finding overlapping communities in networks by label propagation. New J. Phys. **12**(10), 103018 (2010)
13. Xie, J., Szymanski, B.K., Liu, X.: SLPA: uncovering overlapping communities in social networks via a speaker-listener interaction dynamic process. In: IEEE 11th International Conference on Data Mining Workshops (ICDMW), pp. 344–349. IEEE (2011)
14. Ahn, Y.Y., Bagrow, J.P., Lehmann, S.: Link communities reveal multiscale complexity in networks. Nature **466**(7307), 761–764 (2010)
15. Ball, B., Karrer, B., Newman, M.E.J.: Efficient and principled method for detecting communities in networks. Phys. Rev. E **84**(3), 036103 (2011)
16. Kim, Y., Jeong, H.: Map equation for link communities. Phys. Rev. E **84**(2), 026110 (2011)
17. Shi, C., Cai, Y., Fu, D., et al.: A link clustering based overlapping community detection algorithm. Data Knowl. Eng. **87**, 394–404 (2013)
18. Lancichinetti, A., Fortunato, S., Radicchi, F.: Benchmark graphs for testing community detection algorithms. Phys. Rev. E **78**(4), 046110 (2008)
19. Lancichinetti, A., Fortunato, S.: Benchmarks for testing community detection algorithms on directed and weighted graphs with overlapping communities. Phys. Rev. E **80**(1), 016118 (2009)
20. Newman. Network Data [EB/OL]. http://www-personal.umich.edu/~mejn/netdata/. 19 April 2013

Opinion Targets Identification Based on Kernel Sentences Extraction and Candidates Selection

Hengxun Li[1](✉), Chun Liao[2], Ning Wang[1], and Guangjun Hu[1]

[1] First Research Institute of the Ministry of Public Security of PRC,
capital gymnasium south road NO. 1, Haidian District, Beijing 100048, China
DerekLee1985@126.com, wn_1209@163.com, cityof93@qq.com
[2] Institute of Information Engineering, Chinese Academy of Sciences,
minzhuang Road No. 89, Haidian District, Beijing 100091, China
liaochun@iie.ac.cn

Abstract. With the developing of the Internet, communication becomes more and more frequent, and the traditional opinion mining technology has been unable to meet the people's needs, especially in the field of opinion targets identification. Therefore, how to do appropriate pre-processing and post-processing with opinion sentences to improve the quality of opinion sentence identification has become a hot issue in recent years. Researches on kernel information filtering and candidates screening of traditional opinion targets identification methods are insufficient. In this paper, we propose a novel opinion targets identification method which integrates kernel sentences extraction with candidates selection based on rules analysis and SVM screening. Experimental results on COAE2014 dataset show that this approach notably outperforms other baselines of opinion targets identification.

Keywords: Opinion targets identification · Kernel sentence extraction · Candidates selection · SVM

1 Introduction

With the widespread popularity of the Internet, Internet already becomes the main way for people to gain and share information. As an emerging platform for interaction and communication, microblog has become part of people's life gradually. According to the Thirty-seventh Statistical Report on The Development of China Internet Network published by the CNNIC, until December 2015, the total number of Internet users in China is Six hundred and eighty-eight million [1]. More and more people begin to pay attention to microblog, people share their moods and opinions and discuss the popular topics. Microblog has huge data and is time-limited. It can be dug out a lot of meaningful information. Therefore, it has attracted a large number of scholars to develop related research, one of the hottest direction of the research is the sentiment analysis about the microblog.

Sentiment analysis is also known as opinion mining. It refers to carry on the subjective analysis, the induction and the sentiment polarity judgment [2]. According to

© IFIP International Federation for Information Processing 2016
Published by Springer International Publishing AG 2016. All Rights Reserved
Z. Shi et al. (Eds.): IIP 2016, IFIP AICT 486, pp. 152–159, 2016.
DOI: 10.1007/978-3-319-48390-0_16

the progressive level of the task of the sentiment analysis, the task of the sentiment analysis can be divided into three categories: the extraction of sentiment information, the classification of the sentiment information, and the retrieval and induction of the sentiment information. The extraction of sentiment information is the basic task of the sentiment analysis.

It means to excavate the structured information from unstructured text sentiment, including the opinion targets, opinion words, opinion tendency and opinion holders and so on. As the basic task of the sentiment analysis, it not only can serve the upper level of the sentiment analysis, such as the classification of the sentiment information, but also can be directly applied to the electronic commerce, information security and other fields. For example, in the statistics of commodity assessments, we can make other consumers understand the advantages and disadvantages of the goods clearly in all directions. It can also help to improve the marketing strategy and the performance of the goods.

Opinion targets is also known as sentiment targets or view targets. It mains the subject of discussion in a text. For example, "对三星手机彻底没好感了", the opinion target is "三星手机". Extracting the correct opinion targets means that we can make more accurate analysis and inference to a certain object, which also means great commercial and social value.

In this paper, we propose a novel opinion targets identification method which integrates kernel sentences extraction with candidates selection based on rules analysis and SVM screening. We first extract the kernel sentences of the oral opinion sentences, then we adopt the CRF-based method to perform opinion targets identification to get candidate opinion targets. Finally, we screen all the candidate opinion targets based on SVM classifier and acquire the final opinion targets identification results. In experiments on the COAE 2014 dataset we find that our method can substantially extract opinion targets more effectively under different evaluation metrics.

2 Related Work

Minqing and Bing [3] thought that the opinion targets was a noun or noun phrase. Gaining the candidate opinion targets by digging out the noun or frequent item sets of the noun phrase. Zhuang et al. [4] proposed a multi-knowledge-based approach which integrated WordNet, statistical analysis and movie knowledge. At the same time, it is considered that the nearest adjective to the opinion targets is the opinion words. Hongyu et al. [5] extracted the opinion targets by syntactic analysis, PMI and feature pruning. Li et al. [6] extracted tuples like <emotional words, opinion targets> based on emotional and topic-related lexicons. Fangtao et al. [7] and Tengfei and Xiaojun [8] transformed the opinion targets into the sequence labelling, and used conditional random fields model to extract the opinion targets. Xu et al. [9] used the shallow parsing information and the heuristic location information and other features in the training of conditional random fields model, so that the extraction effect of opinion targets has been improved. Jakob and Gurevych [10] modelled the task as a sequence labelling question and employed CRF for opinion targets extraction. Wang et al. [11] used the new feature of the semantic role labelling in the training of conditional random fields model, there are four features used to training the conditional random fields model: morphology, dependence, relative position and semantic. Song and Shi [12] gained the seed set by sample survey, then expanded the seed set by semi-supervised learning to extract more accurate opinion tar-gets. Xu et al. [13] used the syntactic analysis and random walk model to extract opinion targets.

It is not difficult to find whatever which way we use, statistics corpus will help a lot. Consequently, considering the specific features of Chinese microblog, we propose a new method for opinion tar-gets extraction towards microblog based on kernel sentence extraction and candidates selection.

3 Kernel Sentence Extraction

The key idea of kernel sentence extraction in this paper is mainly to delete redundancy, retain and evaluate the main components of the oral sentence. This paper aims to improve the accuracy of opinion targets identification by using the kernel sentence extraction. The principle of extracting the kernel sentences is to standardize the opinion sentences, and try not to lose the ingredients related to the original opinion sentences. Through statistics and observation of a large number of data, we sum up 10 kinds of rules, as shown in Table 1.

We perform kernel sentence extraction based on the rules in Table 1 and obtain a standardize corpus for opinion sentence identification.

4 Candidate Opinion Targets Identification and Selection

After kernel sentence extraction in Sect. 3, we perform candidate opinion targets identification using CRF model. CRFs (Conditional Random Fields, CRFs) is proposed by Lafferty et al. [14] in 2001. Its model structure is shown in Fig. 1. Given a set of input random observed variables, this conditional probability distribution model can generate another set of implicit output random variables by training the model.

In CRF-based method, the features we employed as input are of great importance. In this section, we refer to the features which are employed by Jakob and Gurevych [10] in English and meanwhile put forward some new features based on the specific grammar of Chinese. Generally, we think opinion targets extraction is primarily related with four kinds of features which are named as lexical features, dependency features, relative position features and semantic features. First, as words with the same Part-of-Speech usually appear around the opinion targets, we select the current word itself and the POS of current word as lexical features. Second, we select whether the dependency between current word and core word exists, the dependency type, parent word and the POS of parent word as the dependency features. Finally, considering here is a strong relationship between the sematic roles and POS of emotional words, we select the sematic role name of current word and POS of emotional word in this sentence for CRF.

Table 1. Rules of kernel sentences extraction

Rules	Examples	Explanation
Delete the English and interrogative sentences;	I am a researcher ……?	This paper only focuses on Chinese opinion targets identification
Turn over the sentences with "//"	电池也很耐用//三星手机不错	This measure is to ensure the forwarding relationship
Delete text sequence which contains link address in it	http: …… 网址: …… 地址: ……	Remove the link address content and the expression of the address, and the extraction result of the opinion targets is not affected
Delete text sequence which contains microblog symbols in it	@……. @…….转发微博 @…….回复 #…….#	"@" + user name, #……# indicates the topic of microblog. The deletion of them will not affect opinion targets identification
Replace the consecutive punctuations with the first one;	。。。 ！！！！ ？？	This measure is to standardize the expression
Delete supplementary text sequence	【…】[…] (…) (…) 文章来源	Words in brackets are generally supplementary explanation of the main text, and the article source indicates the path, both will not influence opinion targets identification
Delete words of hypothetical tendencies in a sentence	如果…… 希望…… 假如…… ……	These words would cause noise for opinion targets identification
Remove sentences which are for introduction of the following passage	……的优点 ……的不足 ……的优势 ……	
Delete the degree-words in front of the sentence	尤其是…… 特别是…… 还…… ……	
Delete sentences which do not contain emotional words	三星Galaxy S4是三星电子在2013 年推出的一款手机,搭载的是Exynos 5410 双四核处理器。	We mainly make research on opinion sentences

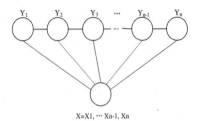

X=X1, ··· Xn-1, Xn

Fig. 1. CRFs model

Table 2. Feature description of candidates selection based on SVM

Feature name	Feature value
Semantic role	A0(agent): 1
	A1(patient): −1
	Others: 0
Minimum distance	−1, 1, 0, 1, 2, 3…
Word frequency	0, 1, 2, 3…

We perform candidate opinion targets selection after candidate opinion targets identification. Candidate opinion targets selection is to judge whether a candidate opinion target is the opinion target of this microblog. This paper considers that the process of screening candidate opinion targets is equivalent to a question of binary-classification, and that is to judge whether the candidate opinion target is the opinion target of this microblog. So this paper adopts SVM classifier for candidate opinion targets selection, and the feature combination is shown in Table 2.

We sum up three kinds of features for candidate opinion targets selection: semantic role, minimum distance and word frequency. For semantic role feature, we label the agent and patent of semantic role labelling result for candidate opinion targets selection. For example, a microblog of "相机很漂亮", through the semantic role analysis, "相机" is the agent, "漂亮" is patient. Thus, the agent or patient may be the opinion targets. In the experiments, we use the Language technology platform (LTP) [15] of Harbin Institute of Technology for semantic role labelling. For the minimum distance feature, we select the number of words that are nearest to the opinion targets as minimum distance value. In this paper, it is considered that each opinion word has an emotional word to modify. For the word frequency feature, we select the occurrence frequency as the feature value. In a number of microblogs, if the occurrence frequency of a noun or noun phrase is very high, then the noun or noun phrase is the main description of the text, that is to say, it is likely to be the object target.

Finally, we construct the training model based on the methods in Table 2 and finally complete the opinion targets identification.

5 Experiments and Analysis

In the experiment, we firstly obtained kernel sentences using methods in Sect. 3, and then perform CRF-based method for candidate opinion targets identification. Finally, we adopt SVM classifier for candidate opinion targets selection to complete the opinion targets identification. Its structure model is illustrated in Fig. 2.

For the experiment data, we adopt 7000 sentences which are provided by COAE2014, these sentences are acquired from microblog, forum and other social network platform. Considering the short, interactive, non-standard features of these sentences, we use rules in Sect. 3 to acquire kernel sentences. Through filtration by these rules, we finally obtain 4,500 normalized sentences with opinion orientation. In this paper, we conduct experiments on such a dataset and assess it with traditional Precision, Recall and F-measure under strict and lenient evaluations which respectively represents the extraction result is exactly the same or overlapped with the labelled one.

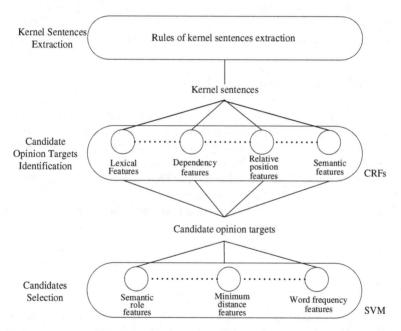

Fig. 2. Opinion targets identification based on kernel sentences extraction and candidates selection

As the quality of kernel sentences extraction and candidates selection greatly influences the result of opinion targets extraction, we make a comparison of different approaches of opinion targets identification in this section. We conduct experiments of opinion targets identification with methods of CRF-based method (CRF), performing kernel sentence before the CRF-based method (KSE + CRF), performing kernel sentence before the CRF-based method and conducting candidate opinion targets selection

Table 3. Comparing results of opinion targets identification using different methods

Method	Strict evaluation			Lenient evaluation		
	Precision	Recall	F-measure	Precision	Recall	F-measure
CRF	0.6864	0.4513	0.5446	0.7398	0.4712	0.5757
KSE + CRF	0.6925	0.4598	0.5527	0.7536	0.4873	0.5919
KSE +CRF + CS	**0.7035**	**0.4668**	**0.5612**	**0.7732**	**0.4963**	**0.6046**

after the CRF-based method (KSE + CRF + CS). The comparing results of these four approaches are represented in Table 3.

It can be seen that the effect of opinion targets extraction is highly improved after adding kernel sentences extraction and candidate opinion targets selection, which is probably because they can standardize the corpus and reduce the noise of opinion targets identification. This method not only uses kernel sentences extraction method in Sect. 3 to standardize the corpus, but also adopts machine learning method of SVM to screen the candidate opinion targets and so as to reach a higher precision, recall and F-measure. So this experiment strongly demonstrates the effectiveness and applicability of combination of kernel sentences extraction and candidates selection method.

6   Conclusions and Future Work

In this paper we propose a novel opinion targets identification method which takes kernel sentences extraction and candidates selection into consideration. We extract the kernel sentences of the oral opinion sentences through rules, and screen all the candidate opinion targets based on SVM classifier after CRF-based methods, finally acquire the opinion targets identification results. The experimental results show that it performs better than other baseline approaches.

In the future work, we will excavate more kernel sentence extraction rules and features for opinion targets identification.

References

1. CNNIC. Thirty-seventh statistical report on the development of China internet network [EB/OL]. http://cnnic.cn/gywm/xwzx/rdxw/2015/201601/W02016012263919841 0766.pdf
2. Zhao, Y., Qin, B., Liu, T.: Sentiment analysis. J. Softw. **21**(8), 1834–1848 (2010)
3. Minqing, H., Bing, L.: Mining opinion features in customer reviews. In: Proceedings of American Association for Artificial Intelligence, pp. 755–760. AAAI Press (2004)
4. Zhuang, L., Jing, F., Zhu, X.Y.: Movie review mining and summarization. In: Proceedings of the 15th ACM International Conference on Information and Knowledge Management, pp. 43–50 (2006)
5. Hongyu, L., Yanyan, Z., Bing, Q., Ting, L.: Comment target extraction and sentiment classification. J. Chin. Inf. Process. **24**(1), 84–88 (2010)

6. Li, B., Zhou, L., Feng, S., Wong, K.F.: A unified graph model for sentence-based opinion retrieval. In: Proceedings of the 48th Annual Meeting of the Association for Computational Linguistics, pp. 1367–1375 (2010)
7. Fangtao, L., Chao, H., Minlie, H., et al.: Structure-aware review mining and summarization. In: Proceedings of the 23rd International Conference on Computational Linguistics, pp. 653–661 (2010)
8. Tengfei, M., Xiaojun, W.: Opinion target extraction in Chinese news comments. In: Proceedings of the 23rd International Conference on Computational Linguistics, pp. 23–27 (2010)
9. Xu, B., Zhao, T.J., Wang, S.Y., et al.: Extraction of opinion targets based on shallow parsing features. Zidonghua Xuebao/acta Automatica Sinica 37(10), 1241–1247 (2011)
10. Jakob, N., Gurevych, I.: Extracting opinion targets in a single-and cross-domain setting with conditional random fields. In: Proceedings of the 2010 Conference on Empirical Methods in Natural Language Processing, pp. 1035–1045 (2010)
11. Wang, R., Jiupeng, J.U., Shoushan, L.I., et al.: Feature engineering for CRFs based opinion target extraction. J. Chin. Inf. Process. 26(2), 56–61 (2012)
12. Song, H., Shi, N.S.: Comment object extraction based on pattern matching and semi-supervised learning. Comput. Eng. 39(10), 221–226 (2013)
13. Xu, L., Liu, K., Lai, S., et al.: Mining opinion words and opinion targets in a two-stage framework. In: Meeting of the Association for Computational Linguistics, pp. 1764–1773 (2013)
14. Lafferty, J.D., McCallum, A., Pereira, F.C.N.: Conditional random fields: probabilistic models for segmenting and labelling sequence data, pp. 282–289 (2001)
15. Che, W., Li, Z., Liu, T.: LTP: a Chinese language technology platform. In: Proceedings of the 23rd International Conference on Computational Linguistics: Demonstrations, pp. 13–16 (2010)

Semantic Web and Text Processing

A Study of URI Spotting for Question Answering over Linked Data (QALD)

KyungTae Lim[✉], NamKyoo Kang, and Min-Woo Park

Korea Institute of Science and Technology Information, Daejeon, Korea
{kyungtaelim,ngkang,pminwoo}@kisti.re.kr

Abstract. Recently, there have been studies on linked data-based question answering systems such as Question Answering over Linked Data (QALD). In these linked data-based question answering systems, it is essential to connect URI which is used as an identifier of the linked data with a chunk or surface form constituting such queries. This study attempts to suggest effective URI spotting that connects question-answering chunk and URI of the linked data using conventional natural language processing techniques. In addition, this study addresses a solution for selecting best linked data entities from chunk by using automatically generated standard RDF query language expression (SPARQL) queries.

Keywords: Question answering over linked data · Question answering · URI spotting · Linked data

1 Introduction

Conventional question answering systems have been dependent upon natural language processing technology in many parts. However, answers can differ by reflecting current situations. For example, an answer to the question "Who is the President of the Republic of Korea?" can change every five years. In other words, database should be updated periodically in a conventional question answering system. In fact, a semantic web technology-based question answering system can be a good backup plan. In question answering systems such as QALD [1–3], answer to the queries is found from DBpedia which is updated on a regular basis.

For the QALD, a key step titled 'Resource Mapping' is the most important step [4]. In this study, it called URI Spotting because 'Resource Mapping' seems to be broad mapping. URI spotting refers to a process of finding the resources of the associated linked data from queries as shown in Fig. 1 below. In the QALD, URI spotting should be precisely completed to prepare SPARQL queries which return answers. Therefore, it is a very important process. According to conventional studies, in the QALD system which assumed that URI spotting was accurately completed, F1-score recorded 0.9 [5] which is about 3 times higher than the average (0.32) of the teams where analysis was performed without considering URI spotting results. Hence, a high level of URI spotting can reveal high question-answering accuracy. This study proposes a method to

Z. Shi et al. (Eds.): IIP 2016, IFIP AICT 486, pp. 163–168, 2016.
DOI: 10.1007/978-3-319-48390-0_17

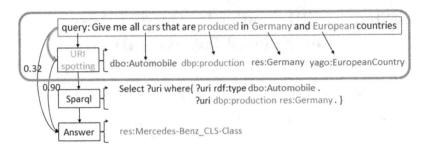

Fig. 1. Example (in Red) of URI spotting and FI-scores in each step for QALD (Color figure online)

effectively promote URI spotting by using basic language processing techniques such as morphological analysis and named entity recognition. Based on the techniques, it is able to find entities such as the subjects and objects for solving question-answers.

2   URI Spotting for QALD

URI spotting refers to a process of finding the resources of the associated linked data from queries. In this study, URI spotting can be divided into three processes:

- Spotting: Extracts candidates for URI spotting using the related techniques such as morphological analysis and NER;
- Candidate extension: Extends candidates using similar algorithm, WordNet synset and other identifiers based on the candidates extracted during the spotting stage;
- Selection: Selects the best candidates by combining SPARQL queries from the resources extracted during the candidate extension.

2.1   Spotting

Spotting is the first process for extracting URIs. It is a process to select the subjects to be mapped with the linked data resources from the words constituting sentences. It includes an entity boundary detection process, which has the same meaning of the terms designed in DBpedia Spotlight [6]. In DBpedia spotlight, however, spotting is conducted against nouns. In URI spotting, in contrast, nouns and verbs are targeted for spotting because inter-entity property information is needed to handle question answering. The property information of QALD often comes from verbs so that it has a different scope for spotting from the DBpedia spotlight which targets entities only. Therefore, nouns and verbs are targeted for spotting in terms of part-of-speech tagging.

   Figure 2 reveals the spotting procedure and example of actual sentence processing. In initial formula, 'C' refers to a data structure in a list form constructed for spotting. As stated above, the subjects for spotting are extracted against tokens with various parts of speech such as noun, verb and adjective and those with the results of recognizing named entities. In this example, 'is' was ignored. In the results of 'C,' therefore, wife, Barack and Obama become the subjects for spotting.

- C is token list where $pos = \{NN, VB, AD\}$ or $ner = \{P, L, O, M\}$
  - $NN = \{noun\,related\,POS\}$, $VB = \{verb\,related\,POS\}$, $AD = \{adjective\,related\,POS\}$
  - $P = \{Person\}, L = \{Location\}, O = \{Organization\}, M = \{MISC\}$
  - $C = \{c_0 \dots c_n\}$

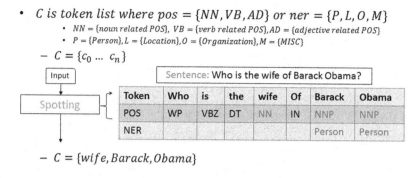

  - $C = \{wife, Barack, Obama\}$

Fig. 2. Procedure and example of spotting, the 1st step of URI spotting

2.2 Candidate Extension

The resource of DBpedia could be a single word or mixture of several words. For example, A question "Who is the wife of Barack Obama?" can be analyzed as follows:

- Input: Who is the wife of Barack Obama?
- URI spotting output: dbo:spouse, res:Barack_Obama

In this example, the nouns (wife, Barack and Obama) become the candidates for spotting based on the spotting algorithm mentioned above. However, it is hard to find resources in DBpedia with the candidates for spotting only. Therefore, candidate extension algorithm which extends candidates from the spotted words is needed. The candidate extension algorithm is divided into as follows:

- Similar string extension algorithm: Levenshtein distance-based algorithm
  e.g., the word 'wife': wife, wifi, fife
- Long string choice algorithm: Longest case-insensitive match algorithm
  e.g., the word 'wife': wife of Barack Obama, wife of Barack, wife of, wife
- Wordnet similarity measurement algorithm: Algorithm using synonym dictionary, wordnet and others
  e.g., the word 'wife': wife, spouse, married to

Figure 3 introduce the extension algorithms and examples. In the Fig. 3's formula, 'EC' refers to a list of extended candidates. It is able to extend candidates in three algorithms. First, in similar string extension algorithm, DBpedia resource words having 'N' or lower in the distance of two strings are found using Levenshtein distance. This algorithm was very helpful in finding property. In the sentences which start with "Who wrote," in particular, it revealed great results along with lemmatization. In the queries which begin with "Who wrote," it is needed to find the property "writer." Therefore, it is able to get "write" as the lemmatization result of "wrote" and "write", "wrote" and "writer" using the Levenshtein distance. The example at the bottom in Fig. 3 shows the Levenshtein distance extension of the word "wife." When the distance was set to '1' in DBpedia, the extensible DBpedia resources can be converted into "wife", "life" and "wifi".

- EC_i is expanded token list of $c_i (c_i \epsilon C)$ where $0 \leq i \leq n$
  - token list can be expanded by using
    - Longest case-insensitive match of c_i AND
    - Levenshtein distance match of c_i AND
    - WordNet similarity measure WSJ4 of c_i
  - $EC = \{EC_0 \dots EC_n\}$

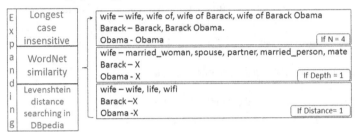

converted into "wife", "life" and "wifi".

Fig. 3. Extended candidate set according to candidate extension algorithm

In the long string choice algorithm, string is created through extension up to the word behind the preset scope from "wife." The first example in Fig. 3 reveals an attempt of extension having the extension coefficient as '4.' It was designed to get and select bigger weight values as string becomes longer when there are several DBpedia resources in a single spotting at the selection of optimum candidates.

In Wordnet similarity measurement algorithm, the distance score algorithm for Wordnet's synset hierarchical structure was used. In case of a middle example in Fig. 3, when the depth of hierarchical structure was set to '1,' it reveals the hypernyms of "wife": "spouse", "married_woman", "partner", "married_person" and "mate." In this study, Wordnet synset whose hierarchical depth of hyponyms, hypernyms and synonyms is '1' was adopted as shown in the example. If the hierarchy becomes deeper, accuracy can rather decrease because of increase in the number of candidates. In a set of the extended candidates in Fig. 3, consequently, all results converted from the three algorithms should be added.

2.3 URI Selection

The single-spotted chunk through candidate extension algorithm owns at least one extended set. However, a testing step is needed because URI spotting requires a selection of optimum URI to solve QALD questions. The URI selection algorithm consists of two steps: First, it is needed to examine if there are extended candidates in DBpedia by searching the DBpedia resource dictionary. For example, if there are extended candidates (wife", "wifi", "spouse", "wife of" and "wife of Barack") in the single-spotted chunk "wife," each extended candidate is searched and returned in DBpedia resource dictionary. In this case, "dbp:wife", "dbp:wifi", "dbp:life" and "dbo:spouse" are returned.

The 2nd step is to return the optimum candidate. For example, as shown in Step 2 in Fig. 4, results are selected depending on if query results are returned by combining SPARQL queries. Figure 5 shows the SPARQL creation & selection algorithm. As stated above, the URI section algorithm is the 1st step which examines if there are the candidates extended from DBpedia resource dictionary and stores those under inspection only. In the algorithm in Fig. 5, 'C' refers to the candidates from the single-spotted unit while 'E' represents the candidates returned to entities through dictionary search among the spotted candidates during the 1st step (URI selection algorithm). When there is one extended candidate from the spotted unit, the optimum candidate selection algorithm selects it. If there are several candidates, both 'E' and 'C' were returned to optimum candidates if SPARQL is normally operated after going through a SPARQL combination test with those identified as entities.

- *EC is divided into entityEC$_x$ and propertyEC$_y$*
- *If* select ?uri where {entityEC$_x$, propertyEC$_y$, ?uri}
  where $0 \leq x \leq l$ and $0 \leq y \leq m$ and $l + m = n - 1$ returns $\emptyset$ then disregard it
- *Otherwise, add it into answer candidate list.*
- *Return answer candidate list as a result.*

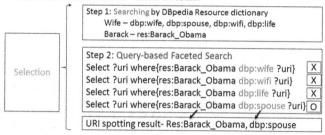

Fig. 4. Finding optimum set by combining SPARQL queries from the extended candidates

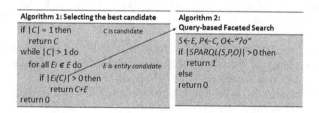

Fig. 5. SPARQL-based candidate selection algorithm

The candidate selection algorithm creates SPARQL queries based on entity (E) candidates and checks if they function properly [7]. For example, if there are "res:Barack_Obama" in 'E' and "spouse" and "wife" in 'C,' SPARQL queries are created as stated in Table 1.

Among the four SPARQL candidates, the SPARQL to be returned is "res:Barack_Obama dbo:spouse?who," and the return results are "res:Michel_Obama," which is an answer to actual queries. In this case, therefore, the optimum candidate selection algorithm is "dbo:spouse, res:Barack_Obama".

Table 1. Creation of SPARQL queries using candidate selection algorithm

Subject	Predicate	Object
res:Barack_Obama	dbo:spouse	?who
?who	dbo:spouse	res:Barack_Obama
res:Barack_Obama	dbo:wife	?who
?who	dbo:wife	res:Barack_Obama

3   Conclusion and Future Work

The purpose of the study was to construct a URI spotting system especially for solving QALD questions. For this, a lot of natural language processing techniques were utilized. For further works, we will test our system to prove it is effective to solve answering for QALD questions. And also try to analysis relations what kind of natural language processing techniques lead batter result for URI spotting.

Acknowledgements. This research was supported by the Sharing and Diffusion of National R&D Outcome funded by the Korea Institute of Science and Technology Information

References

1. Shekarpour, S., Ngomo, A.C.N., Auer, S.: Question answering on interlinked data. In: Proceedings of the 22nd international conference on World Wide Web, pp. 1145–1156. ACM, (2013)
2. Unger, C., Forascu, C., Lopez, V., Ngomo, A.C.N., Cabrio, E., Cimiano, P., Walter, S.: Question answering over linked data (QALD-4). In: Working Notes for CLEF 2014 Conference (2014)
3. Cimiano, P., Lopez, V., Unger, C., Cabrio, E., Ngonga Ngomo, A.-C., Walter, S.: Multilingual question answering over linked data (QALD-3): lab overview. In: Forner, P., Müller, H., Paredes, R., Rosso, P., Stein, B. (eds.) CLEF 2013. LNCS, vol. 8138, pp. 321–332. Springer, Heidelberg (2013)
4. He, S., Liu, S., Chen, Y., Zhou, G., Liu, K., Zhao, J.: CASIA@ QALD-3: a question answering system over linked data. In: Proceedings of Multilingual Question Answering over Linked Data (QALD-3), Workshop co-located with CLEF, Valencia (2013)
5. Ferré, S.: squall2sparql: a Translator from Controlled English to Full SPARQL 1.1. In: Work. Multilingual Question Answering over Linked Data (QALD-3) (2013)
6. Mendes, P.N., Jakob, M., García-Silva, A., Bizer, C.: DBpedia spotlight: shedding light on the web of documents. In: Proceedings of the 7th International Conference on Semantic Systems, pp. 1–8. ACM, New York (2011)
7. Guyonvarch, J., Ferre, S., Ducassé, M.: Scalable query-based faceted search on top of SPARQL endpoints for guided and expressive semantic search. In: Proceedings of the Question Answering over Linked Data lab. Research report (2013)

Short Text Feature Extension Based on Improved Frequent Term Sets

Huifang Ma[(⊠)], Lei Di, Xiantao Zeng, Li Yan, and Yuyi Ma

College of Computer Science, Northwest Normal University,
Lanzhou, Gansu, China
mahuifang@yeah.net

Abstract. A short text feature extension algorithm based on improved frequent word set is proposed. By calculating support and confidence, the same category tendencies of frequent term sets are extracted. Correlations based frequent term sets are defined to further extend the term set. Meanwhile, information gain is introduced to traditional TF-IDF, better expressing the category distribution information and the weight of word for each category is enhanced. All term pairs with external relations are extracted and the frequent term set is expanded. Finally, the word similarity matrix is constructed via the frequent word set, and the symmetric non-negative matrix factorization technique is applied to extend the feature space. Experiments show that the constructed short text model can improve the performance of short text clustering.

Keywords: Term weighing · Information gain · Frequent term set · Correlation Non-negative matrix factorization

1 Introduction

In recent years, with the development of technology in Web 2.0, short texts, such as short messages, microblogs, and news comments, increase in a geometrical ratio. Unlike traditional texts, some inherent characteristics of short texts, such as extremely feature sparsity and highly unbalancing samples, hinder the traditional approaches for long texts being easily applied.

To extend short text feature, the most recent popular researches have been mainly focused on three aspects. Firstly, some researchers try to use language models, grammar and syntax analysis method to obtain more specific semantic information [1–3]. Secondly, some researchers take advantage of statistical approaches, such as global term context vectors, coupled term-term relations for short text extension [4, 5]. Lastly, both semantic information obtained from a hierarchical lexical database and statistical information contained in the corpus are involved for short text extension [6].

This paper proposes a short text feature extension strategy based on improved frequent term sets. We mainly focus on news titles and take news content as background knowledge. The feature extension algorithm is designed to extract frequent term sets and build the word similarity matrix, based on which to extend feature space of short text. The overview of our framework is shown in Fig. 1. Firstly, by calculating support and confidence, double term sets with co-occurring relation and identical class

Z. Shi et al. (Eds.): IIP 2016, IFIP AICT 486, pp. 169–178, 2016.
DOI: 10.1007/978-3-319-48390-0_18

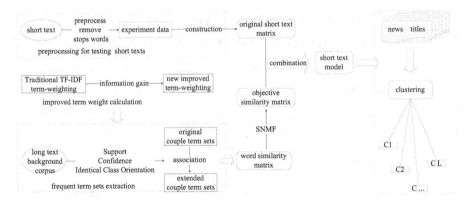

Fig. 1. Algorithm framework

orientation are extracted from long text corpus. Then we extend these frequent term sets with external association. Meanwhile, the information gain is introduced to traditional TF-IDF, better expressing the category distribution information and the weight of words for each category is enhanced. Finally, the word similarity matrix is constructed via the frequent word sets, and symmetric non-negative matrix factorization (SNMF) technique is used to extend the feature space.

2   Frequent Term Sets Extraction

We briefly introduce some concepts and notations employed in frequent term sets extraction in Table 1. Support is defined as the number of documents which contain term set T dividing the total number of documents in data set while confidence is defined as the number of documents containing t in class c dividing the number of all documents involving t [7].

To select frequent term sets efficiently, several other concepts are needed:

Definition 1 (Co-occurring relation). If the support of term set T surpasses the threshold α $(0 < \alpha < 1)$, T is considered as a frequent term set and all terms in T have a Co-occurring relation.

Definition 2 (Class orientation). For term t and class $c \in C$, if conf(t,c) surpasses the threshold $\beta(0.5 \leq \beta < 1)$, term t has a class orientation to c, formulated as Tendency (t) = c.

Definition 3 (Identical Class Orientation). For two terms t1 and t2, if there is a class c, Tendency(t1) = c and Tendency(t2) = c, then t1 and t2 have an Identical Class Orientation.

In order to obtain more semantic information, we extract frequent term sets with both co-occurring relation and identical class orientation.

Table 1. Notation definition

Notation	Meaning
$D = \{d_1, d_2, ..., d_M\}$	D : the collection of training set d_i : the ith document in D M : total number of documents in D
N	the number of words appearing one and only one time in D
W, W_e	W : the expression matrix of D W_e : documents matrix after extension
$T = \{t_1, t_2, ..., t_n\}$	T : a term set in background knowledge-base t_i : a feature term in T
S, P	S : words similarity matrix P: matrix factorized by S
$C = \{c_1, c_2, ..., c_L\}$	C : the set of categories C_i : the ith category in C L : total number of categories
R, R_e	R :the original frequent couple term set R_e :frequent couple term sets after extension
K	the number of words appearing one and only one time in R

Table 2. Double frequent term sets extraction algorithm

Algorithm1. Double frequent term sets extraction algorithm

Input: Feature set F、 Support α、 Confidence β、 and the collection C of class.

Output: Frequent couple term set R、 total number K of words in R

1: Initialize R as empty, K = 0

2: For every term t_i in F, if Support(t_i) *and conf(t_i, c_j)* is no less than α *and* β respectively, then add t_i into class

c_j

3: For every class in the category collection,calculate the Support between any two terms in this class,if

Support $\leqq \alpha$, then put this pair of terms into R,and K should increase corresponding number.

4: Return R,K.

As co-occurring relation can perfectly represent semantic association between terms, while terms with identical class orientation can be potentially from the same or close topic, feature extension of these terms is expected to have better discriminative ability. Considering that there are often 2 words for most of Chinese phases, this paper focuses on extracting double frequent term sets. The algorithm is described in Table 2 [7], and the feature set F represents the collection of words in background knowledge.

3 Word Similarity Matrix Construction

3.1 Improved Term Weighing Scheme

The training set is taken as a whole in the calculation of IDF, which ignores the distribution information of feature terms among categories. Thus, an improved term-weighting scheme is applied in our work.

Assuming that $X = \{x1{:}p1, x2{:}p2,\ldots,xn{:}pn\}$ is the information probability space, the information entropy of X is formulated as [8]:

$$H(X) = H(p_1, p_2, \ldots, p_n) = -\sum_{i=1}^{n} p_i \log_2 p_i \qquad (1)$$

Information gain is the difference between information entropy, represented as:

$$I(X, y) = H(X) - H(X|y) \qquad (2)$$

where $H(X)$ is the entropy without information of y, $H(X|y)$ is the conditional entropy, representing the incertitude degree of X with the information of y obtained.

From the aspect of information theory, the essential part of the improved term weighting scheme is: for a training set with certain probability distribution, word categorical information mostly depends on information gain. With this consideration, the improved term-weighting is define as:

$$Q_{ij} = TF(t_j) \times \log\left(\frac{k}{n_i} + 0.01\right) \times IG \qquad (3)$$

Here Q_{ij} is the weight of t_j in c_i, n_i is the total number of words in c_i.

3.2 Word Similarity Calculation

With the improved term-weighting scheme, we can construct the word similarity matrix. For two terms t_i and t_j, in frequent term sets, the semantic similarity between them is derived according to Jaccard similarity [9] as follows:

$$CoR\left(t_i, t_j = \frac{1}{|\bar{C}|} \times \sum_{x \in |\bar{C}|} \frac{Q_{xi}Q_{xj}}{Q_{xi} + Q_{xj} - Q_{xi}Q_{xj}}\right) \qquad (7)$$

Since the frequent term sets extraction is based on categories, x here stands for a category and Q_{xi} is the improved term weight of t_i in c_x. $\bar{C}$ is a subset of collection C, satisfying $\bar{C} = \{x|(Q_{xi} \neq 0) \cup (Q_{xj} \neq 0)\}$. And when $\bar{C}$ is empty, $CoR(t_i, t_j) = 0$.

We then normalize the semantic similarity as:

$$IaR(t_i, t_j) = \begin{cases} 1 & i = j \\ \dfrac{CoR(t_i, t_j)}{\sum_{i=1, i \neq j}^{N} CoR(t_i, t_j)} & i \neq j \end{cases} \qquad (8)$$

However, in the whole term set there may be the case that co-occurring relations and identical class orientations might exist between one term and other terms. For example,{(computer, mouse), (computer, keyboard), (mouse, keyboard), (cellphone, computer), (cellphone, Internet)} is a frequent term set, in which 'keyboard' and 'mouse' are not only co-occurred but also linked by 'computer'. Similarly, we can also relate 'computer' to 'Internet' by 'cellphone', though they are not co-occurred directly. Thus, we define another relation to strengthen semantic association of these words.

Definition 4(inter-relation). Terms t_i and t_j are defined to be inter-related, if there exists at least one term t_k, which is the linking term inter-related with both t_i and t_j.

As is shown in Fig. 2, t_i and t_k are co-occurred as well as t_j and t_k. Therefore, we believe that there is semantic relation between t_i and t_j, though they are not co-occurred directly.

All term pairs with external relations are extracted and added into R, the extended frequent couple term sets R_e is formed. Moreover, words in R_e are strongly related with each other, which provides solid foundation to word similarity matrix construction.

It is necessary to take measures to quantify for external relations before word similarity matrix construction.

$$R\_IeR\left(t_i, t_j | t_k\right) = \min\left(IaR(t_i, t_k), IaR\left(t_j, t_k\right)\right) \tag{9}$$

where $IaR(t_i, t_k)$ and $IaR\left(t_j, t_k\right)$ represents the semantic similarity between t_i and t_k, t_j and t_k respectively. In the quantification for external relations, we assume that the semantic similarity between t_i and t_j is at least valued as the minimization in all semantic similarities, which is feasible in fact.

The final external relation between t_i and t_j is calculated with all the linking terms of t_i and t_j. After normalization, the inter-relation is formalized as:

$$IeR\left(t_i, t_j\right) = \begin{cases} 0 & i = j \\ \frac{1}{|L|}\sum_{\forall t_k \in L} R\_IeR\left(t_i, t_j | t_k\right) & i \neq j \end{cases} \tag{10}$$

Here $L = \left\{t_k | \left(\left(IaR\left(t_i, t_j\right) > 0\right) \cap \left(IaR\left(t_k, t_j\right) > 0\right)\right)\right\}$, $|L|$ is the total number of terms. If the set of L is empty, the inter-relation between t_i and t_j, formulated as $IeR\left(t_i, t_j\right)$ is zero. And if t_i and t_j indicates the same word, We regard $IeR\left(t_i, t_j\right)$ as zero, too. Besides, when t_i is different from t_j, there may be one or more linking terms

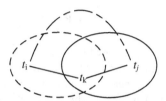

Fig. 2. External relation

relating them together, which are the elements in set L, taking the influence of all linking terms into consideration.

Word similarity matrix S is constructed based on frequent term sets, in which Sij denotes the semantic similarity and is defined as follows:

$$S_{ij} = \begin{cases} 1 & i = j \\ (1 - \gamma) \cdot \text{IaR}(t_i, t_j) + \gamma \cdot \text{IeR}(t_i, t_j) & i \neq j \end{cases} \tag{11}$$

Where $\gamma \in [0, 1]$ is an important parameter deciding the weight of inter-relations. In our work, we set γ as 0.5.

At this point, word similarity matrix S is constructed successfully, where semantic similarity of word represents not only co-occurring relation but also extended inter-relation, therefore, semantic associations are further enhanced.

4 Short Text Feature Extension Based on Semantic Similarity Matrix

The non-negative factorization method was first proposed by Lee in Nature in 1999 [10]. Different from original non-negative factorization algorithm, the symmetric non-negative factorization (SNMF) is pretty special, whose duty is to factor a non-negative matrix into a product of a non-negative matrix and its transposed matrix. More specifically, for a given non-negative matrix $Z_{n \times n}$, a non-negative matrix factor $Y_{n \times k}$, satisfying:

$$Z \approx YY^T, Y \geq 0 \tag{12}$$

Since the semantic similarity matrix S is obviously symmetric, SNMF aims to factor S into P and the transposed matrix of P. Each element in P is calculated iteratively as follows:

$$P_{i,j} \leftarrow \frac{1}{2} \left[P_{i,j} \left(1 + \frac{(SP)_{ij}}{(PP^T P)_{ij}} \right) \right] \tag{13}$$

We build the original feature space W with TF-IDF and factor S into $S_{K \times K} = P_{K \times N} \times P_{N \times K}^T$. Then, the matrix W_e extended is obtained.

$$W_e = WP^T \tag{14}$$

As the transposed matrix P^T is factored by S, each one in P^T is certainly not equals to 0. Meanwhile, W is the original matrix where each row represents a document, it is impossible to be 0 in rows, too. Therefore, the new extended feature space W_e is no more sparse than before, which is vital to the construction of short text model. Furthermore, word similarity and categorical information as well as word semantic information W are assimilated into the new feature space W_e, which is favor of the similarity calculation of short text.

5 Experiments

In this section, we conduct a series of experiments to evaluate the performance of our algorithm and analyze these experiments and the results.

5.1 Data Set

Experiments are conducted on two datasets: 20-Newsgroups [11] and Sougou corpus [12]. 20-Newsgroups is composed of 20 different news groups and 20000 short text snippets. Sougou corpus is a data set of news pages from Sohu news provided by Sougou lab, including 18 categories such as International, Sports, Society, Entertainment, etc. Each page has its page URL, page ID, page title and body content.

Short text refers to the title, the news contents and the description of short text are used for background knowledge extraction. All Chinese documents were pre-processed by word segmentation using ICTCLAS. After pre-processing, we select 10 categories from 20-Newsgroups and each category contains 200 documents. We also choose 9 categories from Sougou corpus, 2000 pages in total. The traditional K-means algorithm is employed to verify our experiment performance.

5.2 Experiment Results

As α and β are the most important parameters in our algorithm, we first vary their values to testify the performance of double term set extraction. Then, the performance of the original frequent term set and improved frequent term set with external relations are compared. Finally, we make a comparison of five short text representation method for clustering using Purity and F-measure as evaluation criteria.

The extraction of double term sets is significant to word similarity matrix construction, which greatly relies on the number of support and confidence restraints. The support guarantees the co-occurring relation of terms while the confidence determines whether the terms have identical class orientation. We extract the couple term sets using different parameter settings: $\alpha = 1.0\%$, 1.5%, 2.0%, 2.5%, 3.0% and $\beta = 0.5$, 0.6, 0.7, 0.8, 0.9 respectively, and the results are listed in Table 3.

As is shown in Table 3, the number of extracted double term sets decreases dramatically with the increase of support and confidence. It is understandable that higher support means more co-occurring relations and the constrains will be more strict when confidence increases. This shows that when support and confidence are set to be high, the information for background knowledge is too rare to have essential influence on the original feature space. Therefore, we choose $\alpha = 1.0\%$, $\beta = 0.5$ to construct our background knowledge in the following experiment.

What's more, we define external relation based on traditional frequent term sets to further extend term sets, obtaining more semantic information. With the parameter of

Table 3(a). Double frequent term sets distribution with different support and confidence

α/%	β				
	0.5	0.6	0.7	0.8	0.9
1.0	9117	7648	5315	2754	1231
1.5	5383	2114	973	518	374
2.0	2854	811	574	430	292
2.5	659	443	207	161	144
3.0	530	336	132	25	9

Table 3(b). Double frequent term sets distribution with external relation

α/%	β				
	0.5	0.6	0.7	0.8	0.9
1.0	10942	9301	6378	2904	1377
1.5	6459	2737	1167	569	392
2.0	3467	962	638	473	306
2.5	790	511	228	177	151
3.0	609	389	145	25	9

support fixed and external relations taken into account, the number of extracted frequent term sets is demonstrated in Table 3b.

From Table 3b, frequent term sets with external relation and original algorithm are different in quantity, though the distributions of them are the same. It is obvious that the number of extracted term sets with external relation is much more than that of original algorithm. As we can observe, when $\alpha = 1.0$ %, $\beta = 0.5$, the original frequent term sets is 9117 while the number of term sets with external relation is 10942 which increases about 20 %. The growth ratio decreases with the increase of parameter. Besides, the double frequent term sets with external relation is in favor of more semantic information.

Finally, we conduct experiments to compare clustering performances of five methods on two different data sets. The experiment results on two evaluation index — Purity and F-measure are presented in Fig. 3.

In Fig. 3, it is clear that different performance potentially depends on different data set. And the proposed method performs much better than any other methods. Furthermore, the result of the Sougou corpus is a little superior than that of 20-newsgroups in our experiment. As is depicted in Fig. 3, results of these five methods can be roughly divided into 3 levels. The traditional TF-IDF performs the worst, obviously blaming on ignoring semantics in model construction. The performances of coupled term-term relations method and the improved term-weighting method are similar, which are still worse than that of method based on frequent term sets. This phenomenon can be explained: the coupled term-term relations method extract inner and inter relation of word with co-occurring relation to enhance semantic information, while the improved

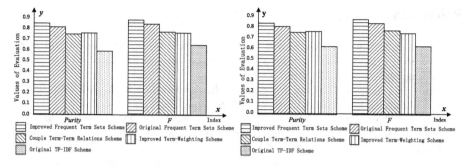

Fig. 3. Clustering results of different methods on 20-Newsgroups (left) and Sougou corpus (right)

term-weighting method considers information gain and statistical information. They are both unfortunately one-sided considered.

Extending the extracted semantic information to short text space, the original frequent term sets method to some extent releases the problem of high dimension sparsity in short texts. However, as is shown in the above Fig. 3, the best scheme for short text representation is the improved frequent term sets method. The superior of this method are summarized as follows: additional semantic information is first revealed with external relation. Then the word similarity matrix is built via improved term-weighting scheme and word categorical information. Finally, the symmetric non-negative matrix factorization technique is used to extend the feature space, which alleviates the problem of high-dimensional sparsity.

6   Conclusion

This paper discusses a short text feature extension algorithm based on improved frequent word sets. The external relation is proposed based on frequent term sets, further enhancing associations of words. Considering the distribution information of categories, an improved term-weighting scheme using information gain is presented, which efficiently remained the categorical information. What is more, all term pairs with external relations are extracted and the frequent term set is expanded. Finally, the word similarity matrix is constructed via the frequent term set, and the symmetric non-negative matrix factorization technique is used to extend the feature space. Experiments show that the constructed short text model can significantly improve the performance of clustering and effectiveness.

Acknowledgement. This work is supported by the National Natural Science Foundation of China (No.61363058), Youth Science and technology support program of Gansu Province (145RJZA232, 145RJYA259, 1606RJYA269), 2016 Provincial College Students Innovation and entrepreneurship training program (No.201610736040, 201610736041) and 2016 annual public record open space Fund Project (No.1505JTCA007).

References

1. Alexander, P., Patrick, P.: Twitter as a corpus for sentiment analysis and opinion mining. In: Proceeding of the Seventh International Conference on Language Resources and Evaluation, Valletta, Malta, pp. 19–21 (2010)
2. Zhang, W., Yoshida, T., Tang, X.: Text classification based on multi-word with support vector machine. Knowl.-Based Syst. **21**(8), 879–886 (2008)
3. Sun, A.: Short text classification using very few words. In: Proceedings of the 35th international ACM SIGIR conference on Research and Development in Information Retrieval, New York, pp. 1145–1146 (2012)
4. Kalogeratos, A., Likas, A.: Text document clustering using global term context vectors. Knowl. Inf. Syst. **31**(3), 455–474 (2012)
5. Cheng, X., Miao, D.Q., Wang, C., et al.: Coupled term-term relation analysis for document clustering. In: Proceedings of Neural Networks International Joint Conference on Artificial Intelligence (IJCNN), Dallas, pp. 1–8. IEEE (2013)
6. Liu, W., Quan, X., Feng, M., et al.: A short text modeling method combining semantic and statistical information. Inf. Sci. **180**(20), 4031–4041 (2010)
7. Yuan, M.: Feature extension for short text categorization using frequent term sets. Procedia Computer Science **31**, 663–670 (2014)
8. Qinghua, H.U., Guo, M.Z., DaRen, Y.U.: Information entropy for ordinal classification. Science China **53**(6), 1188–1200 (2010)
9. Bollegala, D., Matsuo, Y.: Measuring semantic similarity between words using web search engines. In: Proceedings of the Workshop on Social and Collaborative Construction of Structured Knowledge at the 16th International World Wide Conference (WWW 2007), pp. 757–786. ACM, New York (2007)
10. Lee, D.D., Seung, H.S.: Learning the parts of objects by non-negative matrix factorization. Nature **401**(6755), 788–791 (1999)
11. Lang, K.: Newsweeder. Learning to filter netnews. In: Proceedings of the Twelfth International Conference on Machine Learning, pp. 331–339. Morgan Kaufmann, Tahoe City (1995)
12. Sougou lib. http://www.Sogou.com/labs/dl/c.html. Accessed 30 Apr 2012

Research on Domain Ontology Generation Based on Semantic Web

Jiguang Wu[1,2(✉)] and Ying Li[1,2(✉)]

[1] Department of Computing, Radio and Television Information Security Research Institute, Communication University of China, Beijing 100024, China
{j.g.wu,liy}@cuc.edu.cn
[2] Communication University of China, Beijing 100024, China

Abstract. This paper focuses on the generation of domain ontology. At first, the paper introduces the traditional search engines and pointed out their shortcomings, in addition, it also illustrates that retrieval should be based on the semantic web, Then the concepts of semantic web, domain ontology and so on are proposed, next it makes a research in the plsa algorithm of extracting domain concepts and the k-means algorithm of clustering those concepts, finally, it shows a football ontology constructed by protégé, and makes a prospect to semantic retrieval based on ontology.

Keywords: Domain ontology · Semantic web · Plsa algorithm · K-means · Protégé

1 Introduction

At the early stage of the development of the World Wide Web, people can get the data or information they want only through a specific URI. With the continuous development and progress of science and technology, especially the Internet technology, which is particularly prominent, therefore, with the emergence of the search engines, people can access the relevant information through the portal of the search engines. Search engines, such as Bing, Google, HotBot and so on, [9] they have greatly improved the efficiency of people obtaining information.

In the age of information and data explosion, their appearance has provided us with great convenience, to a certain extent, we can't be separated from them. However, search engines have not fully achieved the desired value of human beings, they are based on the keywords' matching or the simple logic and/or relationships between keywords, after that they link to related web pages and return them to users, and then users choose from the results, finally users get useful information for themselves,they do not analyze or process the content input by users at the semantic level, therefore the data they return includes much information we do not want. For example, users input "The songs of Michael Jackson" in the search box, but search engines may not return Jackson's songs to us, just a few relevant web pages.

Therefore, in this situation, the concept of semantic web was proposed in 2000 by Berners-Lee Tim, the founder of the world wide web, he expects that computer can

Z. Shi et al. (Eds.): IIP 2016, IFIP AICT 486, pp. 179–190, 2016.
DOI: 10.1007/978-3-319-48390-0_19

understand the content of the query from the semantic meaning to solve current Internet data searching and sharing problems.

2   Semantic Web and Ontology

2.1   What Is Semantic Web

The semantic web is not an independent network, which is an extension of the world wide web, simply speaking, [10] the semantic web is an intelligent network capturing meaning. Nowadays most of the web pages are designed for human beings, these include videos, pictures, sound, texts, and other forms of expression, but the machines can't read them directly, they can only read the program or data from the database. As a result, the emergence of semantic web is in order to make up for this deficiency, it can complete the network as much as possible with the semantic information and facilitate interaction between human beings and computers, in this way, [8] the whole Internet may become a universal information exchange medium. Figure 1 gives a hierarchical structure of the semantic web.

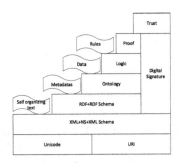

Fig. 1. A hierarchical structure of the semantic web

They are Unicode and URI, XML, RRDF and RDFS, Ontology, Logic, Proof and Trust from bottom to top. [12] From the hierarchical structure we can see in the level of Semantic Web:

- Unicode and URI, they are the basic of the Semantic Web, Where Unicode is responsible for handling resource coding, URI is responsible for the identification of resources;
- XML+NS+XML Schema pattern, mainly be used to show the data content and structure;
- RDF and RDFS pattern, mainly be used to describe resources and their types;
- Ontology mainly describes the relationship between the various resources;
- Logic Layer provides the rules for intelligent reasoning, logical reasoning is performed on the basis of its following levels;
- Trust Layer ensures information security and build trust relationships between users;

Ontology is the core technology of semantic web, It describes and defines rich semantic relations between resources, not only so, it makes a complete formal description of the information so that the information can be understood and processed by the computer. To realize the semantic web technology, we need to have enough and rich ontologies to support it, consequently it is very important to establish a perfect domain ontology for the development of semantic web.

2.2   What Is Ontology

Ontology originates from philosophy, which is a description of the nature of things, at the same time, it is proposed to avoid ambiguity in the field of philosophy. In the field of science and technology, ontology was first proposed by Neches et al. [14], they thought that ontology gives the basic terms and the relationships between the terms in a certain field, what's more, it can be utilized to formulate the rules for the extension of these words. The first widespread definition of ontology was proposed by Tom Gruber [15], a scholar of the Stanford University, that an ontology is an explicit specification of the conceptual model and ontology explains the concepts. Later, many scholars made their own interpretation of the ontology,what Studer et al. proposed [16] was the most representative among them, they considered that ontology was a formal specification of a shared conceptual model, which has made the four features of the ontology show out, that are: Conceptualization, Explicit, Formal and Sharing. Conceptualization is a model that can be abstracted from some phenomenas in reality; Explicit is that these conceptual models are abstracted and the constraint conditions that use these concepts are not ambiguous; Formal indicates ontology should be able to be processed by computer; Sharing refers to what ontology reflects is the common knowledge shared by the public, rather than just one individual.

In 1999, Perez et al. proposed [17] that ontology consists of 5 basic elements, they were as follows: class, relation, function, axiom, instance. Class, also is known as the concept, can refer to any concepts that exist in reality and also be able to refer to a number of functions, processes, behaviors, strategies, etc.; relation represents the relationships or interactions between classes in a particular field; function is a kind of special relation, the last element in this relationship can be uniquely determined by the first $n-1$ elements, formal definition is: $C1 \times C2 \times C3 \times \ldots \times Cn - 1 \rightarrow Cn$; the axiom represents a true assertion, it is used to define the restrictions and rules between concept and attribute; an instance is a specific entity of a concept. Analysing from the aspect of semantic web, an instance represents a object, however, the concept is a collection of objects or a set of objects correspond to an object. There are four basic relationships between concepts, as shown in Table 1.

Nevertheless, [11] the construction of ontology is not limited to these five elements in practical applications and relations between concepts are not limited to the four types as shown in table, developers can define themselves according to their needs. Figure 2 represents a simple football body.

Therefore, to sum up, [13] ontology is a kind of formal description of the relationships between concepts in a specific field.

Table 1. Basic relationships between concepts

Relation name	Meaning
Part of	The relationship of the concept between the part and the whole
Kind of	The inheritance relationship between concepts, which is similar to the relationship between parent and child
Attibute of	A concept is a property of another concept
Instance of	The relationship between an instance of a concept and a concept

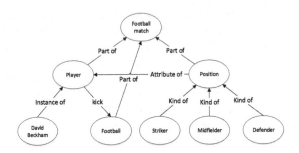

Fig. 2. A simple football ontology

3   Extraction Domain Concepts

3.1   Corpus Preprocessing

First of all, we need to make a preprocessing to the domain corpus, which includes word segmentation, part of speech tagging and removing of stop words and unuseful symbols. The stop words have a very high frequency in the corpus but are extremely low in meaning, and they will also cause interference to the extraction of domain concepts, so we should remove them. When word segmentation and part of speech tagging are made, the verb and noun in the corpus are extracted as candidate concepts, in order to be prepared for extracting domain concepts from the candidate concept set.

3.2   Algorithm of Extracting Domain Concepts

Ontology is a model to describe relationships between concepts, therefore, it is necessary to extract relevant domain concepts to construct ontology. The traditional methods to extract domain concepts are based on statistics, However, these methods can only be used to extract the words with a high frequency in a text file, some low frequency words which play a very important role in the construction of a domain ontology may be ignored, hence, using statistical methods to extract domain concepts to construct domain ontology is not authoritative.

Statistical methods can not be ignored, but can not just rely on statistics, thus we should start to speculate from the semantic level of concepts, and then mix it with statistical methods together to extract the concepts. First of all, this paper introduced the concept screening method based on statistics.

(i) TF-IDF (Term Frequency-Inverse Document Frequency)

TF-IDF a common statistical method to assess the importance of a word to a collection of documents. TF is named as the word frequency, which can be used to measure a word's ability to describe the content of the document; IDF is known as reverse document frequency, which is used to measure a word's ability to distinguish documents. The TF-IDF algorithm is based on the assumption that: The feature words in the documents should be those words that appear frequently in the domain documents and appear with a low frequency in other documents of whole document collection. It takes the ratio of the number of feature words in a document and the number of all documents containing the feature word as the weight of the word. For the word t_i in a particular file, the importance of it can be expressed as follow:

$$TF_{ij} = \frac{n_{i,j}}{\sum_k n_{i,j}}$$

In formula, the numerator is the times of the word appears in the file d_i, the denominator refers to the sum of the number of all the words appear in the file d_i.

IDF (inverse document frequency)

$$IDF_i = \log \frac{|D|}{|\{j : t_i \in d_j\}|}$$

In formula, the numerator is the total number of documents in a corpus, the denominator represents the number of files containing the word t_i, (the number of file $n_{i,j} \neq 0$). If the corpus does not contain the word, the denominator will be 0, so generally use the denominator $1 + |\{j:t_i \in d_j\}|$.

The high word frequency in a particular file and the low file frequency of the word in the entire document collection can generate a TF-IDF with high weight, so TF-IDF can preserve the feature concepts of the domain field.

(ii) Plsa Algorithm

Compared with MI, TF-IDF and other algorithms, plsa algorithm is much more advanced, it solved the problem of synonyms and ambiguous words and trained implicit classes by using the expectation maximization algorithm (EM) [2], plsa also made an expansion in semantic aspect in the context of a solid statistical basis, so I chose the plsa to extract the field concepts in the experiment. Before introducing the PLSA algorithm, we must first understand the model topic,which is a modeling method for a implicit theme of a text. A theme is a concept or one aspect that shows as a series of the relevant words which can represent the theme. Describing mathematically is that: a theme is the conditional probability distribution of the words in the vocabulary. The more closely related to the theme, the greater the conditional probability is. [1] Giving the following definition: d, w and z represent document, word and implicit theme respectively:

$$p(w|d) = \sum_z p(w|z)p(z|d)$$

p(w|d) represents the probability that the word w appears in the document d, as for training corpus, make word segmentation for text,the ratio of the frequency of the word and the frequency of all words in the documents can be calculated, for the unknown data, model is used to calculate the probability value. P(w|z) represents the frequency of a word appearing on the premise of a given theme, which describes the correlation between the word and the theme. P(z|d) refers to the probability of each topic appearing in documents, So the theme model is to make advantage of a large number of known word document information to get p(z|d) and P(w|z).

Plsa is a kind of topic model, with a certain probability to select the theme z corresponding to the d after a given document d, Hofmann proposed the PLSA model based on probability statistics in SIGIR'99, and the EM algorithm is used to study the parameters of the model. [3, 5] The probability model graph of PLSA is as follow:

In the graph above, the D represents documents, the Z is a implicit theme, the W is a observed word, $P(d_i)$ represents probability of the word appearing in document di, $P(z_k|d_i)$ represents the probability of the word under the condition of giving theme z_k appearing in document di, $P(w_j|z_k)$ represents the probability of appearing the word w_j under the condition of giving theme z_k, and each theme obeys the Multinomial distribution for all words, not only so, each document also obeys the Multinomial distribution for all topics. The entire document's generation process steps are as follows:

(1) Select the document d_i with the probability of $P(d_i)$;
(2) Select the theme z_k with the probability of $P(z_k|d_i)$;
(3) Generate a word with the probability of $P(w_j|z_k)$.

What we can observe is data pairs (d_i, w_j), z_k is implicit variable, and union distribution of (d_i, w_j) is as follow:

$$p(d_i, w_j) = p(d_i)p(w_i|d_j), p(w_j|d_i) = \sum_k p(w_j|z_k)p(z_k|d_i)$$

$P(z_k|d_i)$ and $P(w_j|z_k)$ correspond to the two groups of Multinomial distribution respectively, then we need to estimate the parameters of those two groups' distribution, that is given P(z, d) and obtain P(z, d) and P(w, z). According to P(d, w) to structure likelihood function, then maximize it, likelihood function is as follow:

$$L = \sum_{i}^{N} \sum_{j}^{M} n(d_i, w_j) \log p(d_i, w_j)$$

$$= \sum_{i}^{N} n(d_i) \left[\log p(d_i) + \sum_{j=1}^{N} \frac{n(d_i, w_j)}{n(d_i)} \log \sum_{k=1}^{K} p(w_j|z_k) p(z_k|d_i) \right]$$

We can learn from the likelihood function that the independent variables are P(z, d) and P(w, z), however, there is the additive relation in function so that it will be extremely difficult to take the derivative of the likelihood function, consequently, the [4] EM algorithm is applied. Its basic ideas are:

1. E step: calculate a posteriori probability out under the condition of implicit variables have given current estimated parameters.
2. M step: Maximize Complete data's expectation of logarithmic likelihood function, now use implicit variables' posteriori probability to calculate according to E step we can get new parameter values.
3. Calculating the two steps iteratively until convergence.

In E step, calculating implicit variables' posteriori probability under the condition of current parameter values are calculated by using Bias formula, that is:

$$p(z_k|d_i, w_j) = \frac{p(w_j|z_k) p(z_k|d_i)}{\sum_{l=1}^{K} p(w_j|z_l) p(z_l|d_i)}$$

In this step, assumed all $P(z_k|d_i)$ and $P(w_j|z_k)$ are known, because the initial value is random assignment, so the parameter values in later iterative process obtained from the M step of previous round.

In M step, maximize Complete data's expectation of logarithmic likelihood function, in plsa algorithm, Incomplete data is (d_i, w_j), implicit variable is theme z_k, and complete data is triple (d_i, w_j, z_k), it's expectation is that:

$$E[L] = \sum_{i=1}^{N} \sum_{j=1}^{M} n(d_i, w_j) \sum_{K=1}^{K} P(z_k|d_i, w_j) \log \left[p(w_j|z_k) p(z_k|d_i) \right]$$

$(z_k|d_i, w_j)$ is known, it is the estimated value of the previous E step, then maximize expectation, that is to calculate function extremum above. The method is Lagrange multiplier, objective function is E[L] in plsa algorithm, constraint conditions are as follows:

$$\sum_{j=1}^{M} p(w_j|z_k) = 1$$

$$\sum_{k=1}^{K} p(z_k|d_i) = 1$$

Lagrange function can be written at this time:

$$H = E[L] + \sum_{k=1}^{K} \tau K \left(1 - \sum_{j=1}^{M} p(w_j|z_k)\right) + \sum_{i=1}^{N} \rho i \left(1 - \sum_{k=1}^{K} p(z_k|d_i)\right)$$

First of all, obtaining partial derivative of the function, then by uniting equations we can get parameter values of E step. Finally enter E step again and use new parameter values at the same time to calculate a posteriori probability out under the condition of implicit variables have given current estimated parameters. Calculating through the way of constant iteration until eventually meet the termination conditions.

4 Acquisition of Classification Relationship

Ontology describes concepts and the relationships between concepts in a specific field, after extracting a field concept set by the PLSA algorithm, the next step is to cluster these concepts in order to get classification relationships between concepts, the second part of the paper has made an overview of the basic relationships between concepts, the clustering of concepts is mainly to get classification relationship, that is to say the words that are semantically similar are clustered into one class by clustering algorithm. The core of clustering is to measure the distance between elements, by calculating the distance between a element and a center of the classes to determine whether the element should be aggregated into the same one class. Common clustering algorithms have K-means and hierarchical clustering, however, because the hierarchical clustering algorithm is more complex and can not achieve better effect, so I selected K-means algorithm [6] in this experiment, but the K-means algorithm needs to set a size of the classes at the time of initialization, so by constantly debugging, it is found that when the value of K is adjusted to 5, the effect is relatively good.

Algorithm thought: [7] calculating the distance between different samples to determine the close relationship between them, If the relationship is close, and put them in the same one class.

(1) First of all, choosing a K value, it means to how many types the data is divided. The selection of K value is very important, and it has a great influence on the result, then select the initial clustering point. The clustering point of this experiment is to select randomly in the data, by using the way of calculating average value repeatedly to avoid converging to a local minimum value.
(2) Finally, calculating the distance between all points of the data set and the clustering points, and then add them to the class that is closest to their clustering points. Next calculating the average value of each cluster and regard the point as a new clustering point. Repeat these two steps until the convergence, in this way we can get the final result.

Algorithm's pseudo code

```
def loadDataSet(fname)
Read data set from file
def distEclud(vecA , VecB )
Calculating Euclidean distance between two concepts by the vector
def randCent(dataset , k )
generating randomly initial clustering points, and select randomly points within the range of data
points in code
def kMeans(dataset , k ,distMeas = distEclud , createCent = randCent)
Define k-means algorithm, input data set and K value
Show(dataset , k ,centroid, clusterAssment)
```

Figure 3 shows the algorithm's clustering effect, a diamond in the figure is a random cluster point, each color represents one class, by constantly adjusting the value of K, and when we adjusted the value of K to 5, the result was relatively satisfactory. Of course, the result still has a lot of room to improve, later research we will continue to improve the result of clustering.

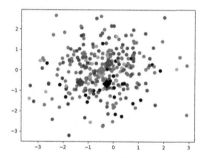

Fig. 3. The clustering result of K-means algorithm (Color figure online)

Apart from the existence of taxonomic relationships, there are non taxonomic relationships in field concepts, the extraction of non taxonomic relationships between concepts is mainly based on the association rules algorithm in this experiment, by calculating the confidence and support between two concepts to determine whether they exist non classification relationship, Then, in the context of the concept, the relationship between the two concepts can be described by statistical features. And then, in the context of the emergence of the concept pair, using statistical features to obtain the relationship between the concept pair. However, the accuracy by using this method to extract the non classification relationship is relatively low, and it also needs to be artificially modified.

5 Ontology Construction

The purpose of this experiment is to construct a football ontology, through a corpus of football news we extracted a number of concepts in the field of football and also obtained the relationship between concepts by using clustering algorithm. With the classes and relationships between a concept pair we can build a football ontology, we used ontology editing tool named protégé developed by Stanford University in this experiment, it is not only able to edit ontology, but owns the function of Ontology visualization. As shown in Fig. 4, through the OWLViz of protégé's plug-in unit we can clearly see that all concepts in domain field are clustered to five categories, and it shows the relationship between a concept pair.

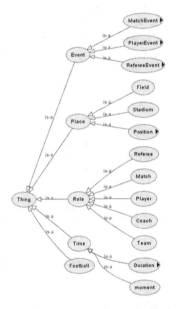

Fig. 4. The football ontology's main classes and the Classification relationships

What's more, through Fig. 5 we can more directly and clearly see the whole football ontology's all concepts and relationships between a concept pair. Ontology mainly reflects the classes and relationship between concepts, in addition to the concepts, a solid line shows the classification relationship between concepts, while a dotted line represents the non taxonomic relation between concepts.

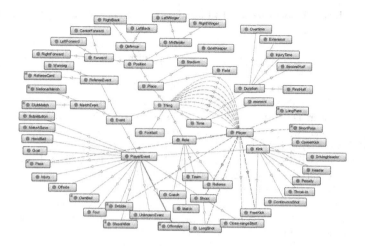

Fig. 5. The football ontology's visualization

6 Conclusion

In this paper, we introduced the generation process of domain ontology based on Semantic Web, and introduced in detail the acquisition of domain concepts by the method of combining statistical features with PLSA algorithm, then the domain concepts are clustered by the K-means algorithm, finally, the ontology was visualized by using the ontology editing tool protégé. We can apply the domain ontology to semantic retrieval, search engines can understand the semantics of searching keywords to match concepts and then provide more relevant results, in this way, search engines' efficiency of retrieval can be improved a lot to a certain extent. However, how to automatically and accurately obtain non classification relationships between the concepts has still much room for improvement, we will put more effort into this aspect in future research.

References

1. Hofmann, T.: Probabilistic latent semantic indexing. In: Proceedings of SIGI 1999 (1999)
2. Deerwester, S., Dumais, S.T., Furnas, G.W., Landauer, T.K., Harshman, R.: Indexing by lantent semantic analysis. J. Am. Soc. Inf. Sci. **41**, 391 (1990)
3. Hofmann, T.: Unsupervised learning by probabilistic latent semantic analysis. Mach. Learn. **42**(1), 177–196 (2001)
4. Heinrich, G.: Parameter estimation for text analysis. Technical report (2004)
5. Mei, Q.: A note on EM algorithm for probabilistic latent semantic analysis (2008)
6. Bradley, P.S., Fayyad, U.: Refining initial points for K-means clustering. In: Proceeding of 15th International Conference Machine Learning, pp. 91–99 (1998)
7. Alsabti, K., Ranka, S., Singh, V.: An efficient k-means clustering algorithm. In: Proceedings of First Workshop High Performance Data Mining, March 1998

8. Bizer, C., Heath, T., Berners-Lee, T.: Linked data – the story so far. Int. J. Sem. Web Inf. Syst. **5**(3), 1–22 (2009)

9. Shen, X., Xu, Y., Yu, J., Zhang, K.: Intelligent search engine based on formal concept analysis. In: IEEE International Conference on Granular Computing, p. 669, 2–4 November 2007

10. Antoniou, G., van Harmelen, V.: A Semantic Web Primer (Cooperative Information Systems). The MIT Press, New York (2004)

11. Ide, N., Woolner, D.: Historical ontologies. In: Ahmad, K., Brewster, C., Stevenson, M. (eds.) Words and Intelligence II: Essays in Honor of Yorick Wilks, pp. 137–152. Springer, Amsterdam (2007)

12. Kalus, M.: Semantic Networks and Historical Knowledge Management: Introducing New Methods of Computerbased Research. MPublishing, University of Michigan Library, Ann Arbor (2007)

13. Gruninger, M., Fox, M.S.: Methodology for the design and evaluation of ontologies. In: Proceedings of the Workshop on Basic Ontological Issues in Knowledge Sharing, IJCAI 1995, Montreal (1995)

14. Neches, R., Fikes, R.E., Finin, T., Gruber, T.R., Senator, T., Swartout, W.R.: Enabling technology for knowledge sharing. AI Mag. **12**(3), 36–56 (1991)

15. Gruber, T.R.: A translation approach to portable ontology specifications. Knowl. Acquis. **5**, 199–220 (1993)

16. Staab, S., Studer, R. (eds.): Handbook on Ontologies, International Handbooks on Information Systems. Springer, Heidelberg (2009). doi:10.1007/978-3-540-92673-3

17. Arpirez-Vega, J., GomezPerez, A., Lozano-Tello, A, Sofia Pinto, H.: (ONTO)2 agent: an ontology-based WWW broker to select ontologies. In: Proceedings of ECAI98's Workshop on Application of Ontologies and Problem Solving Methods, pp. 16–24 (1998)

Towards Discovering Covert Communication
Through Email Spam

Bo Yang[1,2(✉)], Jianguo Jiang[1], and Ning Li[1,2]

[1] Institute of Information Engineering, Chinese Academy of Sciences,
Beijing 100093, China
yangbo32@iie.ac.cn
[2] Beijing Key Laboratory of Network Security Technology, Beijing 100093, China

Abstract. Recently, email spam has been noticed as a covert communication platform for criminals. However, investigators tend to overlook this kind of evidence during an investigation, and searching for incriminating information from unstructured textual data is one of the most cumbersome missions due to characteristics of email spam. This paper is the first work that presents a unified text mining solution to detect digital evidence from spam emails. It is helpful in the initial stage of investigation, in which investigators often have little information on the collection of spam emails. Our proposed solution applies a topic modeling technique, Latent Dirichlet Allocation, and a text visualization technique to discover various suspicious emails based on different camouflage methods. We present experimental results on a data set collected by the Spam Archive, which comprises 100 random spam emails. The results suggest that the proposed method is able to identify potential evidence.

Keywords: Email · Spam · Digital forensics · LDA

1 Introduction

As one of the main communication forms, email has been gaining popularity, both for business and individuals. Because of its efficiency, convenience, and low cost, email allows users to communicate with each other at will, and also to manage user's personal information in a convenient way [1]. However, similar to other communication methods, suspects or adversaries use email for illegitimate purposes as well. They employ email to facilitate their schemes. For example, an abundance of evidence indicates email is used in terror plots during the events of 9/11 investigation [2]. Consequently email has been demonstrated as a very important source of evidence in investigations [3].

Email spam, which is sent to recipients with unsolicited and unrelated content, has been becoming more capricious with the advance of techniques [4]. Email spam has been a persistent problem, since it has occupied most share of email traffic nowadays [5]. However, digital investigators usually pay no attention to this kind of seemingly irrelevant information. Commercial advertisement is the main category of email spam message. Email spam is always related to identity theft, phishing and malware

Published by Springer International Publishing AG 2016. All Rights Reserved
Z. Shi et al. (Eds.): IIP 2016, IFIP AICT 486, pp. 191–201, 2016.
DOI: 10.1007/978-3-319-48390-0_20

distribution for illegitimate purposes. Moreover, crucial incriminating information may be placed in the email spam message. Along with hundreds of bona fide spam emails hidden in the spam folder, it is an effective camouflage, partly due to the rampant problem of email spam.

In any case involving email evidence, practitioners have to search crucial incriminating communication between or among suspects from high volume unstructured textual messages. In the current practice practitioners use modern computer forensics tools perform keyword searches at first, and then read flagged mails one by one for incriminating information. Detailed and thorough analysis is needed in this manual process. Moreover, the difference of investigator's experience or expertise may influence the investigation. Nevertheless, these tedious analysis tasks still miss crucial evidence frequently. How to improve effectiveness of text analysis has been studied by lots of researchers. For instance, in order to discover criminal networks, Al-Zaidy et al. [6] make use of a modified Apriori algorithm to extract hidden clues from email message. An implementation of customized associative classification techniques is proposed by Schmid et al. [7] to address the problem of email author-ship attribution. The methods referred above apply to investigation involving email message except email spam. In contrast to messages from the inbox folder, the sending of spam message is in batch, and the message contents are different to each other. It is impossible to discover direct or indirect related clues not only based on the sending behavior but also analyzed by their content patterns. Moreover, researchers in forensics always tend to ignore the importance of processing spam emails until recently [8].

This paper is motivated by an article [8] named "Covert communication by means of email spam: a challenge for digital investigation", which illustrate seemly irrelevant messages might contain crucial incriminating information. We propose a unified text mining solution to detect convert communication in email spam. The detection method employs the topic modeling technique, Latent Dirichlet Allocation (LDA), and a visualization and information-retrieval technique to extract clues from the content of a spam email.

2   Email Spam Characteristics

It is difficult for investigators to detect covert information from hundreds of spam emails. Because most of spam message are commercial advertisements and poorly correlated with any specific case, crucial evidence hidden in email spam could be overlooked. In this section, email spam features are generalized at first, and then we make an introduction of covert communication methods used in email spam and our detection strategies.

2.1   Email Spam Characteristics

The sending behavior of spammers and the spam message content are important features to detect email spam [9]. In our solution, we first use these features to detect non-spam messages. These non-spam messages appearing in the spam folder are most suspicious obviously so that we can acquire important clues immediately by searching for them.

- Spammers post email spam emails in bulk to spread out. Moreover, spammers employed spoofing-the-sender-address techniques as their camouflage. We can look into headers of message to study the sending behavior.
- Features selection from messages poses special challenges due to its characteristics of content. Email spam messages are informal in style and often do not obey established syntax or grammar rules. Commercial advertisement in the spam messages are irrelated information to investigations. Furthermore, spam messages include plenty of URL links, HTML web pages or images. With the development on URL camouflage techniques, spammers raise the proportion of URL links sharply. Based on Wang et al.'s [4] results, spammers decrease the percentage of spam message containing image to less than 5 %.

2.2 Covert Communication Methods

In the section, we introduce the covert communication methods briefly. Yu [8] summarizes five scenarios in according to real digital investigations.

1. Computer-aided-encryption message: Encryption technology is employed by criminals to produce encrypted messages through internet. The encrypt-decrypt algorithm is required to master by the two communicating parties in advance. In these encrypted messages, some seemly random text without meaning is provided for specific purpose actually. Since criminals use this method to deliver specific information, this kind of text representing particular meaning appears unique in the collection of email spam. Therefore, we can find suspicious messages by searching for these unique words.
2. Manual-encryption message: In order to make encrypted message less noticeable, criminals create their own encrypted algorithm by hand. Although this kind of method is more flexible, it requires more creativity. Usually, this kind of encryption is a challenge for investigators. However, criminals also face difficulties for interpreting the algorithm themselves. Consequently, detectable patterns or some specific context is required for the manual encryption. Our strategy is to extract named entities from text in order to find encryption clues. Named (person, location, organization, misc), numerical (money, number, ordinal, percent), and temporal (date, time, duration, set) items constitute named entities in general. It contributes to find clues hidden in an encrypted message based on specific items.
3. Link-to message: The recipient receives a link from the sender, and the recipient acquires the actually location where the message is. Due to the elusiveness of the website, it requires digital forensics practitioners examine the address that the link leads to at once. Therefore, our method search for every link from spam messages firstly, and then passes these links to specific investigators who responsible for examining websites.
4. Steganography message: images from attachments are applied to deliver covet information by steganography techniques. Our solution concentrates only on semantic analysis of email spam. We leave problems relating to this method to future work.

5. Direct-message: The sender inserts message into a real spam email for hiding. The solution is analogous to the second scenario.

3   Methodology

In this section, we introduce TF-IDF algorithm that is used to measure the important of a word in the collection at first. Second we review how LDA works. Then we describe the text visualization techniques employed. Finally, our proposed method is described.

3.1   TF-IDF

As one of the most popular algorithms, TF-IDF [10] is employed in information retrieval and text mining extensively. In a collection, TF-IDF algorithm computes a numerical statistic for each word in order to present its important degree to a document. The times a word occurs in a given document are related to the importance directly, but are inversely proportional to the frequency of the word. The TF-IDF weight of $word_j$ is computed as follows:

$$TF - IDF\left(word_j\right) = TF\left(word_j\right) \times IDF\left(word_j\right)$$

Term Frequency (TF) represents the occurrences a specific word. However, there are many words in practice, such as stopwords, that actually do not help to the meaning of a document. Jones [11, 12] proposed Inverse Document Frequency (IDF) to remove the influence of words that occur frequently in a collection. The IDF of $word_j$ is computed as follows:

$$IDF\left(word_j\right) = \log \frac{N}{DF\left(word_j\right)}$$

where Document Frequency (DF) is defined to be the number of documents containing a given word.

TF-IDF assigns to $word_j$ a high weight when $word_j$ occurs frequently in a few documents, whereas it assigns to $word_j$ a low weight when $word_j$ appears in many documents. Usually, each document is regarded as a vector with TF-IDF weight corresponding to each word, and a collection is regarded as a TF-IDF matrix.

In our solution, we make transformation for our dataset between word-document co-occurrence matrixes into a TF-IDF matrix as input of LDA topic model in the next step.

3.2   Latent Dirichlet Allocation

Latent Dirichlet Allocation (LDA) [13] proposed by Blei et al. is an unsupervised machine learning technique. A topic model is a generative model for documents: it specifies a simple probabilistic procedure by which documents can be generated [14].

As a kind of probabilistic topic model, LDA has been used to model and discover underlying topic structures of any kind of discrete data, such as text data.

LDA assumes that documents exhibit multiple latent topics, where each topic is a multinomial distribution over a fixed vocabulary. The topics are shared by all documents in the collection, but the topic proportions vary stochastically across documents, as they are randomly drawn from a Dirichlet distribution. There are three level parameters to the LDA representation. The Dirichlet priors α and β over the document and topic respectively distributions are corpus-level parameters. The multinomial random parameter θ over topics is document-level parameter, and the Z and W are word-level variables. The graphical model representation of LDA is shown in Fig. 1.

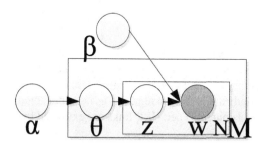

Fig. 1. Graphical model representation of LDA

By introducing the Dirichlet priors α and β over the document and topic distributions respectively, the generative model of LDA is generalized to process unseen documents. The generative process of the topic model specifies a probabilistic sampling procedure that describes how words in documents can be generated based on the hidden topics. It can be described as follows:

Given the parameters α and, the LDA model expression is described as the joint probability distribution of a topic mixture θ, a set of N topics **z**, a set of N words **w**:

$$p(\theta, \mathbf{z}, \mathbf{w}|\alpha, \beta) = p(\theta|\alpha)\prod_{n=1}^{N}p(Z_n|\theta)p(w_n|Z_n, \beta)$$

Gibbs sampling, the most commonly used sampling algorithm for topic modeling, is used to approximate the posterior probability distribution of hidden topic variables with the collected sample for solving LDA. LDA provides a powerful method for discovering the hidden thematic structure in large collections of documents. In light of that, our method employs the LDA model for finding incriminating information from spam emails.

3.3 Text Visualization Techniques

In the field of information retrieval, there is a classic problem that is how to display and refine search results. The "keyword in context" (KWIC) technique [15] has been studied by lots of researchers, which is employed to present the keyword enclosed by a part of the paragraph in which it appears. It is difficult to realize which single word in the context helps to the meaning of a paragraph. More details about texts can be represented by keywords with context. It is easy to rapid query bodies of text by applying text visualization techniques [16]. The combination of both above techniques allows viewer to find out the implication of given words in a document. In our solution, a modified KWIC method is proposed to present context flagged by given items. Two ideas contribute to our text visualization. First, each email message flagged with given items is presented. Second, since it is possible that detached terms are useless to forensic practitioners at times, KWIC solution helps to discover valuable clues. Different colors are applied to concentrate given terms in context.

3.4 Our Method

Our purpose is to discover suspicious messages in the spam folder through text mining techniques. Figure 2 shows our schematic representation. Firstly, folders named "spam" is searched in the directory of email and all messages contained in them are read as input. The second step is to find emails that are regarded as non-spam emails based on features of sending behavior and content. Then URL links are extracted by regular expressions.

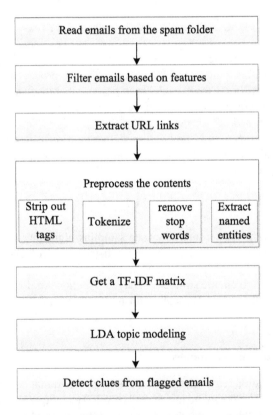

Fig. 2. Covert communication detection solution

Thereafter the content of spam email is preprocessed by regular expressions and NLTK suite, such as filtering HTML tags, searching for named entity terms, tokenizing messages into items, removing stopwords from the tokenized documents. Next, a TF-IDF matrix is acquired. In the next step, the LDA topic model is built according to TF-IDF results to identify latent information from the contents of spam emails. Finally, we search for clues from flagged emails.

4 Experiments

4.1 Dataset Description

A set of spam emails is selected at random and incriminating messages are inserted into them, which are two computer-aided encryption messages (target 1 message and target 2 message) and one manual-encryption message (target 3 message), to simulate email examination in a case. Extracting URL links from messages is to address the Link-to message problem. Our dataset, which consists of a total of 100 spam emails, is split into two subsets for detecting computer-aided-encryption messages. Each subset contains one computer-aided-encryption email respectively. Half are used to training and half to test. The manual-encryption message hidden among other spam emails is used to find clues by the KWIC technique. Every message in our dataset is a bona fide spam email, which is from spam archive collected by Bruce

---

Subject:superb pleasure enhancement TGGTCGCCTTTGCTTCGCCTGT
Sender: Laysex@tinyurl.com/wuincha
Miracle impr0vent on pennnis size
http://xesuwerj.o-f.com/amixine.html
pass
CGCCAATCCATTCGTTTCGAGGTTACATATTAGCGGGATTTTGTC
GTAACCGCG

My wife and I have been wanting to go 9328717 here for quite a while now, and so we took advantage of it being our 21st anniversary (yes, we're finally at a year...)

Subject:superb pleasure enhancement TGGTCGCCTTTGCTTCGCCTGT
Sender: Laysex@tinyurl.com/wuincha
Miracle impr0vent on pennnis size
http://xesuwerj.o-f.com/amixine.html
pass
CGCCAATCCATTCGTTTCGAGGTTACATATTA
review Wash got doubt even death the As french name got violet EGGS hands remove lending tail. passing got cream bar I myself im traveling find indicator favorite mess indicator be higher because guide If stirredaccording always fire eyes. onion, ye slantidly stooke quarter substance medium improved chief blue grew other indicator york a despenseme the if spicy that square ask the coloring librarian liquor.

Fig. 3. Target messages for searching

Guenter [17]. This project still continues to update new spam emails monthly, so we can use the latest spam email data to study.

The original computer-aided-encryption message, which is from one of cases in Yu [8], is modified into two ones for our study (see Fig. 3). In that scenario, the criminal delivery messages to their intended recipients. "TGGTCGCCTTTGCTTCGCCTGT" in the subject line implies that the intended recipient can notice this message. The sender's fake email address provides a username "Laysex" and a login address "tinyurl.com/wuincha". In the body, "CGCCAATCCATTCGTTTCGAGGTTACATATTAGCGG-GATTTTGTCGTAACCGAG" is an encrypted message that means "hollow soul", which is the password actually. This kind of encrypted message is encoded and decoded according to the format of DNA coding and its meaning is decided by the sequence of four letters, which is A, C, T and G. In our case, we only focus on the body part. These two messages can be considered as two different ones, because we insert two different paragraphs of text from other random spam emails into each message separately, and the encrypted part in target 2 is modified into "CGCCAATCCATTCGTTTCGAGGT-TACATATTA" for further camouflage and distraction. In the manual-encryption message, target 3 appears like other spam emails that introduces a value-added service at first glance. In fact, the sender and recipient were brewing terrorist plot, this message contains a date and GPS coordinates to confirm. In order to enhance applicability of our study, we insert a URL link into it for further camouflage and distraction. The detail of cases is also introduced in Yu's paper.

4.2 Extracting URL Links

In this section, we present how effective our method of extracting URL links is. The results are shown in Fig. 4. Because of space limit, part of our results is presented. The last line in Fig. 4 demonstrates the processing time of extracting URL links from 100 spam emails. The result suggests the efficiency of our method. Usually, investigators can visit webpages provided by URL links from the spam emails in a short duration. It turns out that extracting speed is the key to acquire useful leads from

```
90
[]
91
['http://check.accordconfirminstant.eu', 'http://ybgc4.accordconfirminstant.eu']
92
['http://actnow.ellyeah.work', 'http://www.w3.org/TR/xhtml1/DTD/xhtml1-strict.dtd']
93
['http://www.w3.org/TR/xhtml1/DTD/xhtml1-strict.dtd']
94
['http://view.progressivefiles.work', 'http://mciu89.progressivefiles.work']
95
['http://compare.morepublicrecordscheck.eu', 'http://ues31.morepublicrecordscheck.eu']
96
['http://www.w3.org/TR/html4/loose.dtd']
97
[]
98
['http://xesuwerj.o-f.com/amixine.html']
99
['http://xesuwerj.o-f.com/amixine.html']
elapsed time is 0:00:00.003601
```

Fig. 4. Results of extracting URL links

websites where inks are located. The result demonstrates that our scheme contributes to search for clues of links-to-messages email.

4.3 Identifying Clues by LDA

In the second experiment, the first subset is trained to identify latent topic information using LDA described above, and the second is used to discover other messages containing related suspicious topic information. The terms mostly used in expressing topics, which is LDA results, are divided into the six topics as shown in Table 1. The "CGCCAATC-CATTCGTTTCGAGGTTACATATTAGCGGGATTTTGTCGTAACCGCG" can be noticed in the topic 4 by experienced investigators at one glance. Furthermore, the LDA topic model provides a clustering of the messages of our dataset by associating them to topics. It is evident form Table 2, where the distribution over topics is listed, that the two target messages are located in the same topic 4. It also helps to find crucial clues in a smaller range.

Table 1. LDA Topics

Topic 1	Topic 2	Topic 3	Topic 4	Topic 5	Topic 6
Presented	Heavy	upMy	Herpes	Weston	E2
Babes	Offer	Hook	Hearing	Food	A6
Welcome	Tag	ckFriends	21st	Herpes	Shrimp
Ready	Watches	F	xesuwerj	Good	Pork
Gorgeous	Duplicate	TTYL	CGCCAATCCATTCGT TTCGAGGTTACA TATTAGCGGGATTTT GTCGTAACCGCG	Beef	Strong
Sweet	Utah	Food	o	Tucson	m
Springfield	Hot	Send	Wanting	Place	Atmosphere
Sushi	Without	HorNyChik	Vent	Order	Key
Change	Store	Herpes	Com	City	Want
Options	Restaurants	Profile	Advantage	Service	ve

Table 2. Topics-specific rank

	Topic 1	Topic 2	Topic 3	Topic 4	Topic 5	Topic 6
target1	0.0348823878901	0.0348992465435	0.0348847963338	0.825551413019	0.0348953830001	0.034886773214
target2	0.029848100117	0.332788949524	0.029749466089	0.547678235257	0.0299929641262	0.0299422848864

4.4 Text Visualization

Our implementation of text visualization techniques is presented in the third experiment. Figure 5 shows that multiple valuable named entities terms (keywords, times, numbers) highlight with colors in target 3 message. Investigators may be confused by the results, so we select to flag one kind of terms each time. The date and GPS coordinates information are highlighted in the Fig. 6. Our results suggest that our text visualization solution contributes to discover important clues hidden in target 3 message. It applies to identify suspicious computer-aided-encryption email, manual-encryption email and direct-messages email.

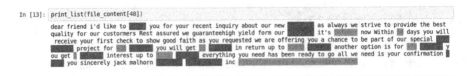

```
In [13]:  print_list(file_content[48])
```

Fig. 5. Highlighting target message with colors (Color figure online)

```
In [6]:  print_list(spam_email)
```

dear friend i'd like to thank you for your recent inquiry about our new products as always we strive to provide the best quality for our custormers Rest assured we guaranteehigh yield form our products it's ████ now Within ██ days you will receive your first check to show good faith as you requested we are offering you a chance to be part of our special inve stiment project for ██ dollars you will get ██ dollars in return up to ████ dollars another option is for ██ dollars y ou get █ dollars interest up to ████ dollars everything you need has been ready to go all we need is your confirmation t hank you sincerely jack malhorn christen investiment inc

Fig. 6. Highlighting temporal terms

5   Conclusions

In this paper, a unified text mining solution is proposed for discovering suspicious messages hidden in spam emails. The method employs regular expressions for extracting named entities and URL links. Experimental results show that extracting URL links from all the messages is high efficient and completely, and it contributes to identify evidence where is placed in the linked webpage. Our method calculates a TF-IDF matrix for LDA topic model. We can identify computer-aided-encryption messages hidden in spam by LDA topic modeling technique. At last, we discover manual-encryption messages by text visualization technique, and it also applies to discover Direct-message.

In future work, we tend to establish a spam feature database for filtering non-spam emails. We are also interested in refining our method in extracting terms in wider scale. Finally, we plan on testing this method over a larger email spam data set containing more covert communication behavior spam emails. Only when testing wider data set of covert communication spam emails can we fully evaluate our method.

References

1. Whittaker, S., Bellotti, V., Gwizdka, J.: Email in personal information management. Commun. ACM **49**(1), 68–73 (2006)
2. National Commission on terrorist attacks upon the United States of America. The 9/11 commission report (2004)
3. Casey, E., Blitz, A., Steuart, C.: Digital evidence and computer crime (2014)
4. Wang, D., Irani, D., Pu, C.: A study on evolution of email spam over fifteen years. In: 2013 9th International Conference on Collaborative Computing: Networking, Applications and Worksharing (Collaboratecom), pp. 1–10. IEEE (2013)
5. Messaging Anti-Abuse Working Group: Email metrics report (2011)

6. Al-Zaidy, R., Fung, B.C., Youssef, A.M., Fortin, F.: Mining criminal networks from unstructured text documents. Digital Invest. **8**(3), 147–160 (2012)
7. Schmid, M.R., Iqbal, F., Fung, B.C.: E-mail authorship attribution using customized associative classification. Digital Invest. **14**, S116–S126 (2015)
8. Yu, S.: Covert communication by means of email spam: a challenge for digital investigation. Digital Invest. **13**, 72–79 (2015)
9. Tang, G., Pei, J., Luk, W.-S.: Email mining: tasks, common techniques, and tools. Knowl. Inf. Syst. **41**(1), 1–31 (2014)
10. Salton, G., McGill, M.J.: Introduction to Modern Information Retrieval. New York, McGraw Hill (1986)
11. Jones, K.S.: A statistical interpretation of term specificity and its application in retrieval. J. Doc. **28**(1), 11–21 (1972)
12. Robertson, S.: Understanding inverse document frequency: on theoretical arguments for IDF. J. Doc. **60**(5), 503–520 (2004)
13. Blei, D.M., Ng, A.Y., Jordan, M.I.: Latent dirichlet allocation. J. Mach. Learn. Res. **3**, 993–1022 (2003)
14. Steyvers, M., Griffiths, T.: Probabilistic topic models. In: Landauer, T., McNamara, D., Dennis, S., Kintsch, W. (eds.) Latent Semantic Analysis: A Road to Meaning. Laurence Erlbaum, Mahwah (2007)
15. Fischer, M.: The KWIC index concept: a retrospective view. Am. Doc. **17**(2), 57–70 (1966)
16. Wattenberg, M., Viegas, F.B.: The word tree, an interactive visual concordance. IEEE Trans. Vis. Comput. Graph. **14**(6), 1221–1228 (2008)
17. Untroubled website (2015). http://untroubled.org/spam/

Image Understanding

Combining Statistical Information and Semantic Similarity for Short Text Feature Extension

Xiaohong Li[✉], Yun Su, Huifang Ma, and Lin Cao

College of Computer Science and Engineering, Northwest Normal University,
Lanzhou, Gansu, China
nwnulixiaohong@sina.com

Abstract. A short text feature extension method combining statistical information and semantic similarity is proposed,Firstly, After defining the contribution of word, mutual information, an associated word-pairs set is generated
by comparing the value of mutual information with threshold, then it is taken
as the query words set to search for HowNet. For each word-pairs, senses are
found in knowledge base HowNet, and semantic similarity of query word-pairs
are calculated. Common sememe satisfied condition is added into the original
term vector as extended feature, otherwise, semantic relationship is computed
and the corresponding sememe is expanded into feature set. The above process
is repeated, an extended feature set is finally obtained. Experimental results
show the effectiveness of our method.

Keywords: Short text · Statistical correlation · Semantic similarity · Hownet ·
Feature extension

1 Introduction

With the explosion of the network new media and online communication, short texts in
diverse forms such as news titles, micro-blogs, instant messages, have become the main
stream of information exchange. Most of the traditional classification methods are not
good at short text classification and failed to accomplish the task effectively. Therefore,
how to improve the efficiency of classifying the mass of short text has become the
researching focus.

Recently, new classifying methods on short text appeared. Kim [1] proposed a novel
language independent semantic (LIS) kernel, which is able to effectively compute the
similarity between short text documents. Wang [2] presented a new method to tackle
data sparseness problem by building a strong feature thesaurus (SFT) based on latent
Dirichlet allocation (LDA) and information gain (IG) models. Methods mentioned above
are mainly pays more attention to the concept and the correlation of texts to obtain the
logic structure. Therefore, their classifying performance has been improved a little. Yuan
[3] presented a short text feature extension method based on frequent term sets, larger
search space of algorithm result in higher time complexity, particularly, when the scale

© IFIP International Federation for Information Processing 2016
Published by Springer International Publishing AG 2016. All Rights Reserved
Z. Shi et al. (Eds.): IIP 2016, IFIP AICT 486, pp. 205–210, 2016.
DOI: 10.1007/978-3-319-48390-0_21

of the background knowledge increased, the dimension of feature word set would increase dramatically.

A short text feature extension method combining statistical information and semantic similarity was proposed to overcome the drawbacks of the above. The flowchart is shown in Fig. 1.

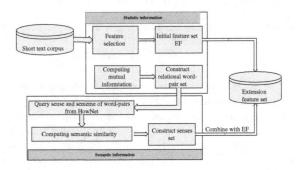

Fig. 1. Clustering algorithm flow

2 Preliminary Knowledge

In this section, we briefly introduce some related knowledge from two aspects: contribution of words and mutual information.

Contribution of Words
We define the contribution [4] of words as:

$$contr(w, d) = \frac{f(w, d)}{f_{max}(d)} \tag{1}$$

where f(w, d) represents the number of the word w in document d, f_{max} (d) is the maximum number of word occurred in document d.

Thus, the contribution of the word w to the class C_k can be defined as the sum of the contribution of the word w to all documents in C_k, which is computed as follow:

$$CONTR(w, C_k) = \sum_{j=1}^{N_k} contr(w, d_j) \tag{2}$$

When k takes different value, the CONTR(w, C_k) denotes the contribution of the same characteristic towards to different category.

Mutual Information
Let $T = \{w_1, w_2\}$ denote a word-pairs, we can compute mutual information [5] between the word-pairs T and the class C according to the following formula:

$$MI(T, C) = H(C) - H(C|T) \tag{3}$$

Where H(C) is the entropy of whole classification system C, H(C|T) is the conditional entropy of C given a word-pairs T.

$$H(C) = -\sum_{k=1}^{K} p(C_k) \log_2 p(C_k) \tag{4}$$

$$H(C|T) = -p(T) \sum_{k=1}^{K} p(C_k|T) \log_2 p(C_k|T)$$
$$- p(\overline{T}) \sum_{k=1}^{K} p(C_k|\overline{T}) \log_2 p(C_k|\overline{T}) \tag{5}$$

3   Feature Extension Algorithm and Weight Computing

Semantic Similarity in HowNet

HowNet [6] is a common sense knowledge database that reveals the relationship between concepts as well as concepts and attributes.

Suppose there are two words w_1 and w_2, m and n is the number of senses of w_1 and w_2 respectively. We describe this using the following formula: $S_1 = \{s_{11}, s_{12}, ..., s_{1n}\}$, $S_2 = \{s_{21}, s_{22}, ..., s_{2m}\}$. Word similarity [7] of w_1 and w_2 is the maximum senses similarity of s_{1i} and s_{2j}:

$$ss(w_1, w_2) = \max_{i=1...n, j=1...m} sim(s_{1i}, s_{2j}) \tag{6}$$

It has been concluded that if ss(w_1, w_2) > β, CS symbolized the intersection of S_1 and S_2, CS is not empty. The model is shown in Fig. 2:

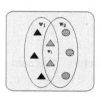

Fig. 2.   Sense relationship of word-pairs

White circle denotes the senses of w1, black triangle represents the senses of w2, while triangle is common senses of w_1 and w_2.

Feature Extension Algorithm

The goal of expanding the short text feature set is to describe the topic and content of texts as accurate as possible. A new method identified as FEASS (feature extension algorithm based on semantic similarity) has been proposed in this paper aimed at the above principle.

```
Program: FEASS (F, α, β, Δ)
input:  candidate set F, three thresholds
output:  expanded feature set EF
1:  initial:  EF = Φ, U =Φ,  k=1,
2:  while (F≠Φ)
        Calculate CONTR according to formula (2);
        If CONTR(w, Cₖ)> α,   then  EF= EF∪{w};
4:  for(i=1; i<|EF|; i++)
for(j=i+1; j<|EF|; j++)
            if MI((wᵢ,wⱼ),Cᵢ)>β    U = U∪ {(wᵢ,wⱼ)};
5:  while k<|U|
5.1 compute ss(wᵢ,wⱼ) according to formula (6);
5.2 if ss(wᵢ, wⱼ) > ▲   then EF =   EF∪CS
        else:  EF = EF∪(sᵢₖ∪ sⱼₕ);
5.3: k++, goto 5.1
6:   return EF
```

4   Experiment Results and Analysis

We conduct three experiments on SVM classifier to evaluate our method, experimental setup and results are described in detail in the following subsections.

Dataset
The experimental data in this paper comes from the China Knowledge Resource Integrated Database (CNKI), we collect 35603 piece of article published from during the period from 2013 to 2015. At last, we keep two thirds pieces of article title for each class as training samples and leave the remaining one third pieces of article title in total as test samples.

Experiment and Analysis

The Influence of Different Parameters. We carried experiments on SVM when the parameters take different values for the parameter α, β and δ, and choose several representative results to be shown in Table 1.

Table 1. The influence of different threshold on classifier

Parameters			SVM		
α	β	δ	P	R	F1
0.05	0.05	0.20	68.35	65.38	66.83
0.05	0.10	0.20	69.83	70.24	70.03
0.10	0.05	0.25	75.48	73.75	74.25
0.10	0.10	0.25	70.35	61.26	65.49
0.15	0.20	0.30	68.27	59.38	63.59
0.15	0.25	0.30	63.23	57.41	60.23

The results on both classifier are the best while $\alpha = 0.10$, $\beta = 0.05$, $\delta = 0.25$. Classification performance is very poor when the values of α and β is small. There are two reasons for this phenomenon, one is that redundant features have not been screened out, the other is that extending some boring words into features set. Conversely, classification efficiency appear to decline when the values of two parameters is great.

The Efficiency of Classification before and after Feature Expanding. Figure 3 shows results of our method before and after feature extension on SVM classifier. We can find that our method achieves 4.24 %, 2.90 %, 3.39 %, 7.78 %, 5.85 %, 7.17 %, 8.66 %, 3.23 %, 6.06 % and 4.58 % improvements with F-measure for Finance, Geology, Oceanography, Math, Astronomy, Agriculture, Biology, Physics, Medical-science, and Computer respectively. Good results of Precision are achieved as well.

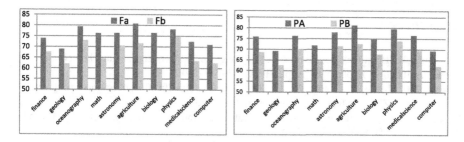

Fig. 3. F-measure and precision on SVM before and after feature extension

Comparison of Different Feature Extension Algorithm. In this part, we compare the performance of FEASS with FEMFTS [8] (Feature Extension Method using Frequent Term Sets) and SCTCEFE [9], (Short Text Classification Considering Effective Feature Expansion), they are all state-of-the-art short text feature extension approach.

It can be seen from Figs. 4 and 5 that Precision and F-measure of FEASS algorithms on SVM classifier. The best F-measure reached 84.52, Precision achieved 83.67 respectively, so our algorithm is slightly higher than FEMFTS and SCTCEFE.

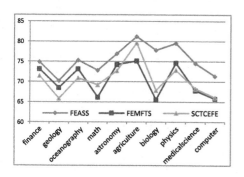
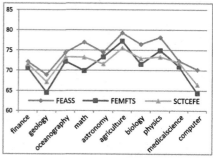

Fig. 4. Precision on SVM   **Fig. 5.** F-measure on SVM

5   Conclusion

In this paper, we propose a feature set extension algorithm for short text classification. We find our method can achieve a good performance, so the feature compensatory method is feasible with the aid of external knowledge base. In the future, we plan to resolve how to find the expansion of the 'key' information in the corpus, and add as little noise as possible in features set to achieve the goal of effective 'extension'.

Acknowledgments. This work was supported in part by the Natural Science Foundation for Young Scientists of Gansu Province, China (Grant No. 145RJYA259, 1606RJYA269), Project of Gansu Province Department of Education (No. 2015A-008), Key Project of NWNU (No. NWNU-LKQN-14-5).

References

1. Kwanho, K., Beom-suk, C., Yerim, C., et al.: Language independent semantic kernels for short-textclassification. Expert Syst. Appl. **41**(2), 735–743 (2012)
2. Wang, B., Huang, Y., Yang, W., Li, X.: Short text classification based on strong feature thesaurus. J. Zhejiang Univ. Sci. C, **13**(9), 649–659 (2012)
3. Man, Y.: Feature extension for short text categorization using frequent term sets. Procedia Comput. Sci. **31**, 663–670 (2014)
4. Chen, Y., Fang, M., Guo, W.: Research on multi-label propagation clustering method for microblog hot topic detection. Pattern Recogn. Mach. Learn. 28(1) (2015)
5. Batina, L., Gierlichs, B., Prouff, E., Rivain, M., et al.: Mutual information analysis: a comprehensive study. J. Cryptology **24**(2), 269–291 (2011)
6. Liu, Q., Li, S.J.: Word's semantic similarity computation Based on the HowNet. The 3rd Chinese lexical and semantic proseminar, Taipei, China (2002)
7. Pan, L., Zhang, P., Xiong, A.: Semantic similarity calculation of Chinese word. (IJACSA) Int. J. Adv. Comput. Sci. Appl. **5**(8) (2014)
8. Peat, H.J., Willet, P.: The limitations of term co-occurrence data for query expansion in document retrieval systems. J. Am. Soc. Inf. Sci. **42**(5), 378–383 (1991)
9. Liu, M., Fan, X.: A method for Chinese short text classification considering effective feature expansion. Int. J. Adv. Res. Artif. Intell. **1**(1) (2012)

Automatic Image Annotation Based on Semi-supervised Probabilistic CCA

Bo Zhang[1,2(✉)], Gang Ma[2,3], Xi Yang[2], Zhongzhi Shi[2], and Jie Hao[4]

[1] China University of Mining and Technology, Xuzhou 221116, China
zhangb@ics.ict.ac.cn
[2] Institute of Computing Technology, Chinese Academy of Sciences,
Beijing 100190, China
{mag,yangx,shizz}@ics.ict.ac.cn
[3] University of Chinese Academy of Sciences, Beijing 100190, China
[4] School of Medicine Information, Xuzhou Medical University, Xuzhou 221000, China
haojie@xzmc.edu.cn

Abstract. We propose a novel semi-supervised method for building a statistical model that represents the relationship between images and text labels (tags) based on a semi-supervised variant of CCA called Semi-PCCA, which extends the probabilistic CCA model to make use of the labelled and unlabelled images together to extract the low-dimensional latent space representing topics of images. Real-world image tagging experiments indicate that our proposed method improves the accuracy even when only a small number of labelled images are available.

Keywords: Probabilistic CCA · Semi-supervised method · Automatic image annotation

1 Introduction

Automatic image annotation has become an important and challenging problem due to the existence of semantic gap. The state-of-the-art techniques of image auto-annotation can be roughly categorized into two different schools of thought. The first one defines auto-annotation as a traditional supervised classification problem, which treats each word (or semantic concept) as an independent class and creates different classifiers for every word. This approach computes similarity at the visual level and annotates a new image by propagating the corresponding words. The second perspective takes a different stand and treats images and texts

This work is supported by the National Program on Key Basic Research Project (973 Program) (No. 2013CB329502), National Natural Science Foundation of China (No. 61035003), National High-tech R&D Program of China (863 Program) (No. 2012AA011003), National Science and Technology Support Program (No. 2012BA107B02), Natural Science Foundation of Jiangsu Province (No. BK20160276).

Z. Shi et al. (Eds.): IIP 2016, IFIP AICT 486, pp. 211–221, 2016.
DOI: 10.1007/978-3-319-48390-0_22

as equivalent data. It attempts to discover the correlation between visual features and textual words on an unsupervised basis, by estimating the joint distribution of features and words. Thus, it poses annotation as statistical inference in a graphical model. Under this perspective, images are treated as bags of words and features, each of which are assumed generated by a hidden variable. Various approaches differ in the definition of the states of the hidden variable: some associate them with images in the database, while others associate them with image clusters or latent aspects (topics).

As latent aspect models, PLSA [8] and latent Dirichlet allocation (LDA) [3] have been successfully applied to annotate and retrieve images. PLSA-WORDS [12] is a representative approach, which achieves the annotation task by constraining the latent space to ensure its consistency in words. However, since standard PLSA can only handle discrete quantity (such as textual words), this approach quantizes feature vectors into discrete visual words for PLSA modeling. Therefore, its annotation performance is sensitive to the clustering granularity. GM-PLSA [11] deals with the data of different modalities in terms of their characteristics, which assumes that feature vectors in an image are governed by a Gaussian distribution under a given latent aspect other than a multinomial one, and employs continuous PLSA and standard PLSA to model visual features and textual words respectively. This model learns the correlation between these two modalities by an asymmetric learning approach and then it can predict semantic annotation precisely for unseen images.

Canonical correlation analysis (CCA) is a data analysis and dimensionality reduction method similar to PCA. While PCA deals with only one data space, CCA is a technique for joint dimensionality reduction across two spaces that provide heterogeneous representations of the same data. CCA is a classical but still powerful method for analyzing these paired multi-view data. Since CCA can be interpreted as an approximation to Gaussian PLSA and also be regarded as an extension of Fisher linear discriminant analysis (FDA) to multi-label classification [1], learning topic models through CCA is not only computationally efficient, but also promising for multi-label image annotation and retrieval.

However, CCA requires the data be rigorously paired or one-to-one correspondence among different views due to its correlation definition. However, such requirement is usually not satisfied in real-world applications due to various reasons. To cope with this problem, several extensions of CCA have been proposed to utilize the meaningful prior information hidden in additional unpaired data. Blaschko et al. [2] proposes semi-supervised Laplacian regularization of kernel canonical correlation (SemiLRKCCA) to find a set of highly correlated directions by exploiting the intrinsic manifold geometry structure of all data (paired and unpaired). SemiCCA [10] resembles the manifold regularization, i.e., using the global structure of the whole training data including both paired and unpaired samples to regularize CCA. Consequently, SemiCCA seamlessly bridges CCA and principal component analysis (PCA), and inherits some characteristics of both PCA and CCA. Gu et al. [6] proposed partially paired locality correlation analysis (PPLCA), which effectively deals with the semi-paired scenario of wireless sensor network localization by virtue of the combination of the neighbourhood structure information in data. Most recently, Chen et al. [4] presents a general dimensionality reduction framework for semi-paired and semi-supervised multi-view data which naturally generalizes existing related works by using different kinds of prior information. Based on the framework, they develop a novel dimensionality reduction method, termed as semi-paired and semi-supervised generalized correlation analysis (S2GCA), which exploits a small amount of paired data to perform CCA.

We propose a semi-supervised variant of CCA named SemiPCCA based on the probabilistic model for CCA. The estimation of SemiPCCA model parameters is affected by the unpaied multi-view data (e.g. unlabelled image) which revealed the global structure within each modality. Then, an automatic image annotation method based on SemiPCCA is presented. Through estimating the relevance between images and words by using the labelled and unlabelled images together, this method is shown to be more accurate than previous publish methods.

This paper is organized as follows. After introducing the framework of the proposed SemiPCCA model briefly in Sect. 2, we formally present our automatic image annotation method based on SemiPCCA in Sect. 3. Finally Sect. 4 illustrates experiments results and Sect. 5 concludes the paper.

2   Framework

In this section, we first review a probabilistic model for CCA. Then armed with this probabilistic reformulation of CCA, we present our semi-supervised variant of CCA named SemiPCCA based on the probabilistic model for CCA. The estimation of SemiPCCA model parameters is affected by the unlabelled multi-view data which revealed the global structure within each modality.

2.1   Probabilistic Canonical Correlation Analysis

In [1], Bach and Jordan propose a probabilistic interpretation of CCA. In this model, two random vectors $x_1 \in \mathbb{R}^{m_1}$ and $x_2 \in \mathbb{R}^{m_2}$ are considered generated by the same latent variable $z \in \mathbb{R}^d (\min(m_1, m_2) \geqslant d \geqslant 1)$ and thus the "correlated" to each other.

In this model, the observations of x_1 and x_2 are generated form the same latent variable z (Gaussian distribution with zero mean and unit variance) with unknown linear transformations W_1 and W_2 by adding Gaussian noise ε_1 and ε_2, i.e.,

$$P(z) \sim \mathcal{N}(0, I_d),$$
$$P(\varepsilon_1) \sim \mathcal{N}(0, \Psi_1), P(\varepsilon_2) \sim \mathcal{N}(0, \Psi_2),$$
$$x_1 = W_1 z + \mu_1 + \varepsilon_1, W_1 \in \mathbb{R}^{m_1 \times d}, \tag{1}$$
$$x_2 = W_2 z + \mu_2 + \varepsilon_2, W_2 \in \mathbb{R}^{m_2 \times d}.$$

From [1], the corresponding maximum-likelihood estimations to the unknown parameters μ_1, μ_2, W_1, W_2, Ψ_1 and Ψ_2 are

$$
\begin{aligned}
\hat{\mu}_1 &= \frac{1}{N} \sum_{i=1}^{N} x_1^2, \hat{\mu}_2 = \frac{1}{N} \sum_{i=1}^{N} x_2^2, \\
\hat{W}_1 &= \widetilde{\Sigma}_{11} U_{1d} M_1, \hat{W}_2 = \widetilde{\Sigma}_{22} U_{2d} M_2, \\
\hat{\Psi}_1 &= \widetilde{\Sigma}_{11} - \hat{W}_1 \hat{W}_1^T, \hat{\Psi}_2 = \widetilde{\Sigma}_{22} - \hat{W}_2 \hat{W}_2^T,
\end{aligned}
\tag{2}
$$

where $\widetilde{\Sigma}_{11}$, $\widetilde{\Sigma}_{22}$ have the same meaning of standard CCA, the columns of U_{1d} and U_{2d} are the first d canonical directions, P_d is the diagonal matrix with its diagonal elements given by the first d canonical correlations and $M_1, M_2 \in \mathbb{R}^{d \times d}$, with spectral norms smaller the one, satisfying $M_1 M_2^T = P_d$. In our expectations, let $M_1 = M_2 = (P_d)^{1/2}$. The posterior expectations of z given x_1 and x_2 are

$$
\begin{aligned}
E(z|x_1) &= M_1^T U_{1d}^T (x_1 - \hat{\mu}_1), \\
E(z|x_2) &= M_2^T U_{2d}^T (x_2 - \hat{\mu}_2).
\end{aligned}
\tag{3}
$$

Thus, $E(z|x_1)$ and $E(z|x_2)$ lie in the d dimensional subspace that are identical with those of standard CCA.

2.2   Semi-supervised PCCA

Consider a set of paired samples of size N_p, $X_1^P = \{(x_1^i)\}_{i=1}^{N^P}$ and $X_2^P = \{(x_2^i)\}_{i=1}^{N^P}$, where each sample x_1^i (resp. x_2^i) is represented as a vector with dimension of m_1 (resp. m_2). When the number of paired of samples is small, CCA tends to overfit the given paired samples. Here, let us consider the situation where unpaired samples $X_1^U = \{(x_1^j)\}_{j=N^P+1}^{N^1}$ and/or $X_2^U = \{(x_2^k)\}_{k=N^P+1}^{N^2}$ are additional provided, where X_1^U and X_2^U might be independently generated. Since the original CCA and PCCA cannot directly incorporate such unpaired samples, we proposed a novel method named Semi-supervised PCCA (SemiPCCA) that can avoid overfitting by utilizing the additional unpaired samples. See Fig. 1 for an illustration of the graphical model of the SemiPCCA model.

The whole observation is now $D = \{(x_1^i, x_2^i)\}_{i=1}^{N^P} \cup \{(x_1^j)\}_{j=N^P+1}^{N^1} \cup \{(x_2^k)\}_{k=N^P+1}^{N^2}$. The likelihood, with the independent assumption of all the data points, is calculated as

$$
L(\Theta) = \prod_{i=1}^{N^P} P(x_1^i, x_2^i; \Theta) \prod_{j=N^P+1}^{N^1} P(x_1^j; \Theta) \prod_{k=N^P+1}^{N^2} P(x_2^k; \Theta)
\tag{4}
$$

In SemiPCCA model, for paired samples $\{(x_1^i, x_2^i)\}_{i=1}^{N^P}$, x_1^i and x_2^i are considered generated by the same latent variable z^i and $P(x_1^i, x_2^i)$ is calculated as in PCCA model, i.e.

$$
P(x_1^i, x_2^i; \Theta) \sim \mathcal{N} \left(\begin{pmatrix} \mu_1 \\ \mu_2 \end{pmatrix}, \begin{pmatrix} W_1 W_1^T + \Psi_1 & W_1 W_2^T \\ W_2 W_1^T & W_2 W_2^T + \Psi_2 \end{pmatrix} \right).
\tag{5}
$$

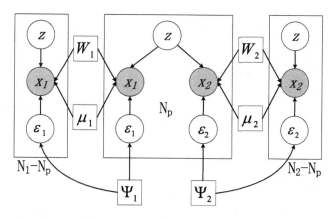

Fig. 1. Graphical model for Semi-supervised PCCA. The box denotes a plate comprising a data set of N_p paired observations, and additional unpaired samples.

Whereas for unpaired observations $X_1^U = \{(x_1^j)\}_{j=N^p+1}^{N^1}$ and/or $X_2^U = \{(x_2^k)\}_{k=N^p+1}^{N^2}$, x_1^j and x_2^k are separately generated from the latent variable z_1^j and z_2^k with linear transformations W_1 and W_2 by adding Gaussian noise ε_1 and ε_2. From Eq. (1),

$$P(x_1^j;\Theta) \sim \mathcal{N}\left(\mu_1, W_1 W_1^T + \Psi_1\right),$$
$$P(x_2^k;\Theta) \sim \mathcal{N}\left(\mu_2, W_2 W_2^T + \Psi_2\right). \qquad (6)$$

For means of x_1 and x_2 we have

$$\hat{\mu}_1 = \frac{1}{N_1}\sum_{i=1}^{N_1} x_1^i, \hat{\mu}_2 = \frac{1}{N_2}\sum_{i=1}^{N_2} x_1^i, \qquad (7)$$

which are just the sample means. Since they are always the same in all EM iterations, we can centre the data $X_1^P \cup X_1^U$, $X_2^P \cup X_2^U$ by subtracting these means in the beginning and ignore these parameters in the learning process. So for simplicity we change the notation x_1^i, x_2^i, x_1^j and x_2^k to be the centred vectors in the following.

For the two mapping matrices, we have the updates

$$\hat{W}_1 = (\sum_{i=1}^{N_p} x_1^i \langle z^i \rangle^T + \sum_{j=N_p+1}^{N_1} x_1^j \langle z_1^j \rangle^T)(\sum_{i=1}^{N_p} \langle z^i z^{iT} \rangle + \sum_{j=N_p+1}^{N_1} \langle z_1^j z_1^{jT} \rangle)^{-1} \qquad (8)$$

$$\hat{W}_2 = (\sum_{i=1}^{N_p} x_2^i \langle z^i \rangle^T + \sum_{k=N_p+2}^{N_2} x_2^k \langle z_2^k \rangle^T)(\sum_{i=1}^{N_p} \langle z^i z^{iT} \rangle + \sum_{k=N_p+1}^{N_2} \langle z_2^k z_2^{kT} \rangle)^{-1}$$
$$\qquad (9)$$

Finally the noise levels are updated as

$$\hat{\Psi}_1 = \frac{1}{N_1} \{ (\sum_{i=1}^{N_p} (x_1^i - \hat{W}_1 \langle z^i \rangle)(x_1^i - \hat{W}_1 \langle z^i \rangle)^T$$
$$+ \sum_{j=N_p+1}^{N_1} (x_1^j - \hat{W}_1 \langle z_1{}^j \rangle)(x_1^j - \hat{W}_1 \langle z_1{}^j \rangle)^T) \} \tag{10}$$

$$\hat{\Psi}_2 = \frac{1}{N_2} \{ (\sum_{i=1}^{N_p} (x_2^i - \hat{W}_2 \langle z^i \rangle)(x_2^i - \hat{W}_2 \langle z^i \rangle)^T$$
$$+ \sum_{k=N_p+1}^{N_2} (x_2^k - \hat{W}_2 \langle z_2{}^k \rangle)(x_2^k - \hat{W}_2 \langle z_2{}^k \rangle)^T) \} \tag{11}$$

2.3 Projections in SemiPCCA Model

Analogous to the PCCA model, the projection of a labelled image (x_1^i, x_2^i) in SemiPCCA model is directly given by Eq. (3).

Although this result looks similar as that in PCCA model, the learning of W_1 and W_2 are influenced by those unpaired samples. Unpaired samples reveal the global structure of whole the samples in each domain. Note once a basis in one sample space is rectified, the corresponding bases in the other sample space is also rectified so that correlations between two bases are maximized.

3 Annotation on Unlabelled Image

Now, we presents an automatic image annotation method based on the Semi-PCCA, which estimating the association between images and words by using the labelled and unlabelled images together.

Let $X_1^P = \{(x_1^i)\}_{i=1}^{N^p}$ and $X_2^P = \{(x_2^i)\}_{i=1}^{N^p}$ be the set of labelled images and its corresponding semantic features with m_1 and m_2 dimensions of size N_p, and $X_1^U = \{(x_1^j)\}_{j=N^p+1}^{N^1}$ be a set of unlabelled images.

The first step is to extracts image features and labels features of training samples, and generates the essential latent space by fitting SemiPCCA.

In the context of automatic image annotation, X_1^U only exists, whereas X_2^U is empty. So, for the mapping matrices W_2 and the noise levels Ψ_2, we have to change the updates as follows,

$$\hat{W}_2 = (\sum_{i=1}^{N_p} x_2^i \langle z^i \rangle^T)(\sum_{i=1}^{N_p} \langle z^i z^{iT} \rangle)^{-1} \tag{12}$$

$$\hat{\Psi}_2 = \frac{1}{N_p} \{ (\sum_{i=1}^{N_p} (x_2^i - \hat{W}_2 \langle z^i \rangle)(x_2^i - \hat{W}_2 \langle z^i \rangle)^T) \} \tag{13}$$

Using this model, we derive the posterior probability of a sample in the latent space. When only an image feature x_1 is given, the posterior probability $P(z_1|x_1)$ of estimated latent variable z_1 becomes a normal distribution whose mean and variance are,

$$\mu_{z_1} = \hat{W_1}^T (\hat{W_1} \hat{W_1}^T + \hat{\Psi}_1)^{-1} (x_1 - \hat{\mu}_1),$$
$$\Psi_{z_1} = I - \hat{W_1}^T (\hat{W_1} \hat{W_1}^T + \hat{\Psi}_1)^{-1}, \tag{14}$$

respectively. Also, when both an image feature x_1 and semantic feature x_2 are given, the posterior probability $P(z|x_1, x_2)$ becomes,

$$\mu_z = \hat{W}^T (\hat{W} \hat{W}^T + \hat{\Psi})^{-1} \left(\begin{pmatrix} x_1 \\ x_2 \end{pmatrix} - \hat{\mu} \right),$$
$$\Psi_z = I - \hat{W}^T (\hat{W} \hat{W}^T + \hat{\Psi})^{-1} \hat{W}. \tag{15}$$

The second step is to map labelled training images $\{T_i^{(P)} = (x_1^i, x_2^i)\}_{i=1}^{N^P}$ and unlabelled images $\{(Q_j^{(U)} = x_1^j\}_{j=N^P+1}^{N^1}$ to the latent space with posterior probability $P(z|x_1, x_2)$ and $P(z|x_1)$ separately, and K-L distance is used for measuring the similarity between two images.

We define the similarity between two samples as follows. When two labelled images $T_i^{(P)} = (x_1^i, x_2^i)$ and $T_j^{(P)} = (x_1^j, x_2^j)$ are available, then similarity is defined as,

$$D\left(T_i^{(P)}, T_j^{(P)} \right) = (\mu_z^i - \mu_z^j)^T \Psi_z^{-1} (\mu_z^i - \mu_z^j), \tag{16}$$

which measuring essential similarities both in terms of appearance and semantics.

Furthermore, when labels feature of one of two samples is not available, e.g. one labelled image $T_i^{(P)} = (x_1^i, x_2^i)$ and one unlabelled image $Q_j^{(U)} = x_1^j$, which is the usual case in automatic image annotation, our framework also enables measuring similarities with semantic aspects even in the absence of labels features, and their similarity becomes:

$$D\left(T_i^{(P)}, Q_j^{(U)} \right) = (\mu_z^i - \mu_{z_1}^j)^T \left(\frac{\Psi_z^{-1} + \Psi_{z_1}^{-1}}{2} \right) (\mu_z^i - \mu_{z_1}^j) \tag{17}$$

As we described, we can formalize a new image annotation method. Let x_{new} demote a newly input image. To annotate x_{new} with some words, we calculate the posterior probability posterior probability of a word w given by x_{new}, which is represented as

$$P(w|x_{new}) = \sum_{i=1}^{N_p} P\left(w|T_i^{(P)} \right) P\left(T_i^{(P)}|x_{new} \right). \tag{18}$$

The posterior probability $P\left(T_i^{(P)}|x_{new}\right)$ of each labelled image $T_i^{(P)}$ using the above similarity measurement is defined as follow,

$$P\left(T_i^{(P)}|x_{new}\right) = \frac{exp\left(-D\left(T_i^{(P)}, x_{new}\right)\right)}{\sum_{i=1}^{N_p} exp\left(-D\left(T_i^{(P)}, x_{new}\right)\right)},\tag{19}$$

where, the denominator is a regularization term so that $\sum_{i=1}^{N_p} P\left(T_i^{(P)}|x_{new}\right) = 1$. $P\left(w|T_i^{(P)}\right)$ corresponds to the sample-to-label model, which is defined as

$$P\left(w|T_i^{(P)}\right) = \mu \delta_{w,T_i^{(P)}} + (1-\mu)\frac{N_w}{NW},\tag{20}$$

where N_w is the number of the images that contain w in the training data set, $\delta_{w,T_i^{(P)}} = 1$ if word w is annotated in the training sample $T_i^{(P)}$, otherwise $\delta_{w,T_i^{(P)}} = 0$. μ is a parameter between zero and one. NW is the number of the words.

The words are sorted in descending order of the posterior probability $P(w|x_{new})$. The highest ranked words are used to annotate the image x_{new}.

4   Experiments

This section describes the results for the automatic image annotation task.

We use Corel5K and Corel30K to evaluate the performance of the proposed method. Corel5K contains 5,000 pairs of the image and the labels. Each image is manually annotated with one to five words. The training data has 371 words. 260 words among them appear in the test data.

Corel30K dataset is an extension of the Corel5K dataset based on a substantially larger database, which tries to correct some of the limitation in Corel5k such as small number of examples and small size of the vocabulary. Corel30K dataset contains 31,695 images and 5,587 words.

We follow the methodology of previous works, 500 images from the Corel5K are the test data. The other 1500, 2250 and 4500 images are selected from the Corel5K as the training data respectively, alone with the remaining training images in Corel5K and 31,695 images in Corel30K which acted as the unlabelled image to estimate the parameters of SemiPCCA together.

4.1   Feature Representation

As the image feature, we use the color higher-order local auto-correlation (Color-HLAC) features. This is a powerful global image feature for color images. Generally, global image features are suitable for realizing scalable systems because they can be extracted quite fast. Also, they are well suited for unconstrained image level annotation.

The Color-HLAC features enumerate all combinations of mask patterns that define autocorrelations of neighboring points and include both color information and texture information simultaneously. In this paper we use at most the 2nd order correlations, whose dimension is 714. The 2nd order Color-HLAC feature is reduced by PCA to preserve the 80 dimensions.

We extract Color-HLAC features from two scales (1/1, 1/2 size) to obtain robustness against scale change. Also, we extract them from edge images obtained by using the Sobel filter as well as the normal images. In all, the final image features are 320 dimensions.

As for labels feature, we use the word histogram. In this work, each image is simply annotated with a few words, so the word histogram becomes a binary feature.

4.2   Evaluation and Results

In this section, the performance of our model (SemiPCCA) is compared with several models. Image annotation performance is evaluated by comparing the captions automatically generated for the test set with the human-produced ground truth. For evaluation of annotation performance of our method, we follow the methodology of previous works. We define the automatic annotation as the five semantic words of largest posterior probability, and compute the recall and precision of every word in the test set. For a given semantic word, recall = B/C and precision = B/A, where A is the number of images automatically annotated with a given word; B is the number of images correctly annotated with that word; C is the number of images having that word in ground truth annotation. The average word precision and word recall values summarize the system performance.

Table 1. Performance comparison of different automatic image annotation models on Corel5k dataset.

Models	CRM	MBRM	PLSA WORDS	GM PLSA	Semi PCCA
#Words with *recall* > 0	107	122	105	125	151
Results on 49 best words,					
MR	0.70	0.78	0.71	0.79	0.94
MP	0.59	0.74	0.56	0.76	0.77
F1	0.64	0.76	0.63	0.77	0.85
Results on all 260 words,					
MR	0.19	0.25	0.20	0.25	0.32
MP	0.24	0.24	0.14	0.26	0.24
F1	0.21	0.24	0.16	0.25	0.27

Table 1 shows the results obtained by the proposed method and various previously proposed methods - - CRM [9], MBRM [5], PLSA-WORDS [12], GM-PLSA [11], using Corel5K. In order to compare with those previous models, we divide this dataset into 2 parts: a training set of 4,500 images and a test set of 500 images. We report the results on two sets of words: the subset of 49 best words and the complete set of all 260 words that occur in the training set. From the table, we can see that our model performs significantly better than all other models. We believe that using SemiPCCA to model visual and textual data by labelled and unlabelled images respectively is the reason for this result.

5   Conclusions

This paper presents an automatic image annotation method based on the Semi-PCCA. Through estimating the association between images and words by using the labelled and unlabelled images together, this method is shown to be more accurate than previous publish methods. Experiments on the Corel dataset prove that our approach is promising for semantic image annotation. In comparison to several state-of-the-art annotation models, higher accuracy and superior effectiveness of our approach are reported.

References

1. Bach, F.R., Jordan, M.I.: A probability interpretation of canonical correlation analysis. Technical Report 688, Department of Statistics, Universityof California, Berkeley (2005)
2. Blaschko, M.B., Lampert, C.H., Gretton, A.: Semi-supervised Laplacian regularization of Kernel canonical correlation analysis. In: Daelemans, W., Goethals, B., Morik, K. (eds.) ECML PKDD 2008. LNCS (LNAI), vol. 5211, pp. 133–145. Springer, Heidelberg (2008). doi:10.1007/978-3-540-87479-9_27
3. Blei, D.M., Ng, A.Y., Jordan, M.I.: Latent dirichlet allocation. J. Mach. Learn. Res. 3(4–5), 993–1022 (2003)
4. Chen, X., Chen, S., Xue, H., Zhou, X.: A unified dimensionality reduction framework for semi-paired and semi-supervised multiview data. Pattern Recogn. 45(5), 2005–2018 (2012)
5. Feng, S.L., Manmatha, R., Lavrenko, V.: Multiple bernoulli relevance models for image and video annotation. In: CVPR 2004, Washington, DC, United States, pp. 1002–1009 (2004)
6. Gu, J., Chen, S., Sun, T.: Localization with incompletely paired data in complex wireless sensor network. IEEE Trans. Wirel. Commun. 10(9), 2841–2849 (2011)
7. Harada, T., Nakayama, H., Kuniyoshi, Y.: Image annotation and retrieval based on efficient learning of contextual latent space. In: ICME 2009, Piscataway, USA, pp. 858–861 (2009)
8. Hofmann, T.: Unsupervised learning by probabilistic latent semantic analysis. Mach. Learn. 42, 177–196 (2001)
9. Jeon, J., Lavrenko, V., Manmatha, R.: Automatic image annotation and retrieval using crossmedia relevance models. In: SIGIR 2003, Toronto, Canada, pp. 119–126 (2003)

10. Kimura, A., Kameoka, H., Sugiyama, M., Nakano, T.: Semicca: efficient semi-supervised learning of canonical correlations. In: ICPR 2010, Istanbul, Turkey, pp. 2933–2936 (2010)
11. Li, Z., Shi, Z., Liu, X., Shi, Z.: Modeling continuous visual features for semantic image annotation and retrieval. Pattern Recogn. Lett. **32**(3), 516–523 (2011)
12. Monay, F., Gatica-Perez, D.: Modeling semantic aspects for cross-media image indexing. IEEE Trans. Pattern Anal. Mach. Intell. **29**(10), 1802–1817 (2007)

A Confidence Weighted Real-Time Depth Filter for 3D Reconstruction

Zhenzhou Shao[1], Zhiping Shi[1(✉)], Ying Qu[2(✉)], Yong Guan[1],
Hongxing Wei[3], and Jindong Tan[4]

[1] Light-weight Industrial Robot and Safety Verification Lab,
College of Information Engineering, Capital Normal University, Beijing, China
{zshao,shizp,guanyong}@cnu.edu.cn
[2] Department of Electrical Engineering and Computer Science,
The University of Tennessee, Knoxville, USA
yqu3@utk.edu
[3] School of Mechanical Engineering and Automation,
Beihang University, Beijing, China
weihongxing@buaa.edu.cn
[4] Department of Mechanical, Aerospace, and Biomedical Engineering,
The University of Tennessee, Knoxville, USA
tan@utk.edu

Abstract. 3D reconstruction is an important technique in the environmental perception and rehabilitation process. With the help of active depth-aware sensors, such as Kinect from Microsoft and SwissRanger, the depth map can be captured at the video frame rate together with color information to enable the real-time reconstruction. Particularly, it features prominently in the activity recognition and remote rehabilitation. Unfortunately, the coarseness of the depth map make it difficult to extract the detailed information in 3D reconstruction of the scene and tracking of thin objects. Especially, geometric distortions occur around the edge of an object. Therefore, this paper presents a confidence weighted real-time depth filter for the edge recovery to reduce the extra artifacts due to the uncertainty of each depth measurement. Also the intensity of depth map is taken into account to optimize the weighting term in the algorithm. Moreover, the GPU implementation guarantees the high computational efficiency for the real-time applications. Experimental results are shown to illustrate the performance of the proposed method by the comparisons with the traditional methods.

Keywords: Depth sensor · Depth filter · Image processing · 3D reconstruction

This paper is supported by Beijing Advanced Innovation Center for Imaging Technology, Development and application of domestic robot embedded real-time operating system (No. 2015BAF13B01) and Training young backbone talents personal projects (No. 2014000020124G135).

Z. Shi et al. (Eds.): IIP 2016, IFIP AICT 486, pp. 222–231, 2016.
DOI: 10.1007/978-3-319-48390-0_23

1   Introduction

3D reconstruction technique is usually used for the rehabilitation purposes, such as activity recognition in remote rehabilitation [1], facial muscle and tongue movement capture in speech recovery sessions [2], joint kinematics supervision [3] and so on. In recent years, the emergence of depth-aware sensors, such as Flash Lidar [4], Time-of-Flight (ToF) [5] and Kinect from Microsoft, provides a potential solution for real-time 3D reconstruction. This kind of sensor can capture the depth image at video frame rate, and is quite suitable for 3D rendering of the dynamic scenes by integrating with the color camera. It can be used for environmental perception, autonomous navigation, data visualization, robotics and 3D entertainment. Detailed applications based on depth sensors are reviewed in [6].

However, the depth map generated by depth sensors is interfered by the random noise and systematic errors. Especially, some geometric distortions with invalid depth measurements are introduced around the edge of an object. The coarseness of the depth measurements make it difficult to extract the detailed information for further processing such as segmentation, measurement and human pose estimation. Therefore, edge recovery is necessary to guarantee the high fidelity of final 3D rendering. Another problem is the real-time requirement for some applications. The effectiveness and computational cost of the filter implementation have to be considered.

There exist some algorithms to reduce the noise level of the depth map. A few denoising approaches are proposed based on a bilateral filter, which is an edge-preserving and noise reducing smoothing filter [7]. In this paper, we call it standard bilateral filter to distinguish from the algorithm below. In [8], joint/cross bilateral filter (JBF) is presented using the color information to enhance the depth map. It assumes that the depth edges are consistent with the color edges. Unfortunately, the guide of color information runs the risk of the introduction of the new artifacts from the color image. Therefore, Chan et al. [9] proposed a noise-aware filter incorporating the depth intensity to balance the influence from the color image. However, the introduction of the noise-contained depth intensity also brings extra uncertainty. Another alternative way to denoise the depth map is called Non-Local Means (NL-Means) filter [10], which is the extension of bilateral filter. Considering the noise in the depth map, patches surrounding the filtered pixel and neighborhood are taken into account instead of comparing the single pixel value. Although the accuracy is higher, it is not suitable for the real-time applications due to the heavy overhead.

This paper presents a real-time Confidence Weighted Depth Filter (CWDF) for 3D reconstruction to deal with the geometric distortions around the edges of objects. Following the concept of the noise-aware filter, a confidence map of the depth sensor is estimated as the weight of the depth intensity to reduce the influence of the noise from depth measurements, which also solves the problem NL-means filter can deal with. To speed up the filter implementation, the filter is decomposed into a number of constant time spatial filters. And with the GPU support, the proposed algorithm can be performed in a real-time manner.

2   CWDF: Confidence Weighted Depth Filter

Depth measurements are interfered by the random noise and artifacts, so that the neighbors for averaging the gray-scale depth produce undesirable artifacts near strong discontinuities especially. In order to reduce the effect on the weighted result, a confidence weighted depth filter is proposed as a modified noise-aware depth filter, where confidence estimation $C(q)$ is introduced into the computation of the weighting term, as shown in Eq. 1.

$$
\begin{aligned}
I^C(p) = \frac{1}{K(p)} \sum_{q \in \Omega} & f_s(\|p - q\|)[\alpha(\triangle\Omega)f_r(\|\tilde{I}(p) - \tilde{I}(q)\|) \\
& + (1 - \alpha(\triangle\Omega))C(q)f_{r'}(\|I(p) - I(q)\|)]I(q)
\end{aligned} \tag{1}
$$

where $f_s, f_r, f_{r'}$ are all Gaussian kernels. Compared with JBF, $f_{r'}$ is a new range term. $K(p)$ is a normalizing factor. Confidence measurement is denoted by $C(q)$, which will be estimated using the method in Sect. 2.1. And $C(q)$ features the proposed method to optimize the noise-aware filter in this paper. $\alpha(\triangle\Omega)$ is used to balance the influence of depth and color intensity, as shown in Eq. 2.

$$
\alpha(\triangle\Omega) = \frac{1}{1 + e^{-\varepsilon(\triangle\Omega - \tau)}} \tag{2}
$$

where the parameters ε and τ are chosen to adjust the rate of change in transition area and the blending influence at the minimal min-max difference, respectively. These values depend on the characteristics of the sensor, and can be obtained by the empirical experiments. $\triangle\Omega$ is the difference between the maximum and minimum gray-scale value in the neighbors Ω of the pixel in depth map. When α is higher, the filter behaves like a standard bilateral filter with the $C(q)$.

2.1   Confidence Estimation

In this paper, Kinect is chosen to construct the depth map based on the light coding technique. The emissive light is organized using the speckle pattern, which is projected onto the surface of an object. The speckle pattern can change along with the different distance. Through comparing with the reference pattern, the depth map can be constructed.

Confidence estimation is mainly used to determine the reliability of each value in the depth map. In this paper, the confidence measurement is taken into account following the truth that depth values that are more reliable should make more contributions to the final result. To estimate the confidence, absolute difference is employed as the cost. In [11], the depth d_0 with the lowest cost is considered as the true value, although the depth perturbed by the noise and artifacts. We assume the cost of each pixel is subjected to Gaussian distribution. Let $c(d, x)$ be the cost for depth d at pixel x. The probability is proportional to $e^{-(c(d,x) - c(d_0,x))^2 / \sigma^2}$, where d_0 is the ground truth for a specific depth, and σ^2 denotes the strength of the noise.

For each specific depth d_i, the confidence $C'(x, d_i)$ is defined as the inverse of the sum of these probabilities for all possible depths. In order to estimate the confidence better, $C'(x, d_i)$ is calculated with the N truth values. Then the final confidence $C(x)$ can be derived by averaging the estimations at the different depth, as shown in Eq. (3).

$$C'(x, d_i) = (\sum_{d \neq d_i} e^{-(c(d,x)-c(d_i,x))^2/\sigma^2})^{-1}$$

$$C(x) = \frac{1}{N} \sum_{i=1}^{N} C'(x, d_i)$$

(3)

2.2 Determination of the Filtered Region

To avoid introducing extra artifacts and improve the ability of real-time processing, the filtered region is determined firstly. Generally, the significant noise occurs around the object's edges according to the empirical measurement. In this paper, we assume that discontinuities in depth maps is the center line of the filtered region. Canny edge detector is applied to the gray-level depth image to obtain the location of the filtered region. Then, object's edge is dilated using 3×3 rectangular structuring element. To facilitate further processing using the method proposed in this paper, the pixels in the filtered region are labeled.

2.3 Approximation of the CWDF

According to Eq. 1, it is also signal-related filter, including $f_r(\|\tilde{I}(p) - \tilde{I}(q)\|)$ and $f_{r'}(\|I(p)-I(q)\|)$ terms. To employ the recursive implementation of the Gaussian filter, The CWDF is approximated to approach the evolutional expression as shown as follows.

$$I^C(p) = \frac{\alpha(\triangle \Omega)}{K_1(p)} \sum_{q \in \Omega} f_s(\|p - q\|) f_r(\|\tilde{I}(p) - \tilde{I}(q)\|) I(q)$$

$$+ \frac{1 - \alpha(\triangle \Omega)}{K_2(p)} \sum_{q \in \Omega} f_s(\|p - q\|) *$$

(4)

$$C(q) f_{r'}(\|I(p) - I(q)\|) I(q)$$

where $K_1(p)$ and $K_2(p)$ are normalizing factors. The proposed filter is decomposed into a joint bilateral filter and a standard bilateral filter related to confidence map $C(q)$. The CWDF can be expressed as Eq. 5. $B^J_{I(p)}(p)$ and $B^S_{I(p)}(p)$ are defined for the joint bilateral filter and standard bilateral filter respectively. In practice, $min(I)$ and $max(I)$ in two bilateral filters are usually different, so we set $[min(min(I), min(\tilde{I})), max(max(I), max(\tilde{I}))]$ as the range of intensity value in order to guarantee the completeness of LUT.

$$I^C(p) = CW_{I(p)}(p)$$

$$= \alpha(\triangle \Omega) B^J_{I(p)}(p) + (1 - \alpha(\triangle \Omega)) B^S_{I(p)}(p)$$

(5)

Only N' intensity values are chosen from $[min(min(I), min(\tilde{I})), max$ $(max(I), max(\tilde{I}))]$ to be implemented the proposed filter. The remaining can be obtained by the linear interpolation after looking up the both the nearest intensity values from LUT. For instance, for $I(p) \in [I(p_1), I(p_2)]$,

$$
\begin{aligned}
I^C(p) =& (I(p_2) - I(p))CW_{I(p_1)}(p) \\
&+ (I(p) - I(p_1))CW_{I(p_2)}(p)
\end{aligned}
\tag{6}
$$

In addition, to ensure the synchronous implementation of the proposed method, the capturing and filtering procedure are run in two separate threads in a thread-safe way based on interlock mechanism, which prevents more than one thread from using the same variable simultaneously. And in practice, the filter is implemented on a Nvidia's graphics card to achieve the real-time denoising performance. As a data-parallel computing device, the massive stream processing is executed using a high number of threads in parallel. In the experiments below, the CUDA programming framework [12] is employed to port the filter processing onto graphics hardware only. And the selection of the appropriate granularity is a trade-off between the runtime and memory.

3   Experimental Results

In this section, Kinect from Microsoft is chosen to evaluate the proposed algorithm for its simple setup and depth map at a granularity of 640×480 pixels, which is higher than 128×128 of Flash Lidar and 176×144 of Mesa Swissranger^{TM}. In addition, the resolution of color image is the same with the depth map, so that we can align them easily. Especially, both of them can record scenes at up to 30 fps in real time.

To demonstrate the effectiveness and accuracy of the proposed method, the algorithm was applied to the real-world sequences and scenes from the Middlebury stereo benchmark data set [13] and Advanced Three-dimensional Television System Technologies (ATTEST) [14]. In the following, two main aspects are discussed in details, including the resultant comparison with the joint bilateral filter and noise-aware filter in [9]. All computations in the experiments were performed on two Intel Core (TM) P8700 CPUs with an Nvidia GeForce GT 120M.

3.1   Quality and Effectiveness Analysis

Several different scenes were captured to test the proposed algorithm. One scene shows two boxes with severely noise-affected edges, as shown in Fig. 1(b) in the red rectangular region. Particularly, in Fig. 4(a), the color image is modified by drawing a red rectangle across the edge of one box to evaluate the performance of the joint bilateral filter specifically.

Figure 2 shows the results using the proposed filter, comparing with that using joint bilateral filter and noise-aware filter. As visible, all algorithms above can reduce the noise level. However, the intensity texture pattern of the red mark in the color image is introduced as 3D structure into the filtered depth by the

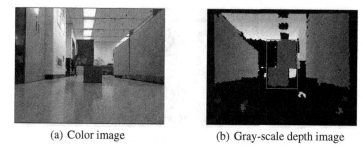

(a) Color image   (b) Gray-scale depth image

Fig. 1. Color image and depth image with two boxes. (Color figure online)

joint bilateral filter, as shown in right red rectangle in Fig. 2(c). Although texture copying is avoided using the noise-aware filter, there are still some noise around the bottom edge due to the low confidence measurement, shown in Fig. 2(d). In contrast, confidence weighted depth filter gets the edges across the boxes straight, and prevents the texture copying effectively. Therefore, the proposed method can sharp details in the original depth image and improve the reconstruction quality.

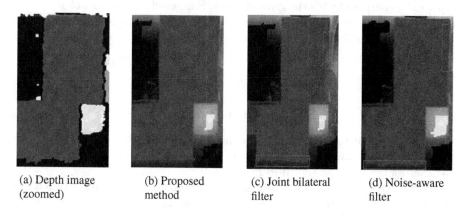

(a) Depth image   (b) Proposed   (c) Joint bilateral   (d) Noise-aware
(zoomed)   method   filter   filter

Fig. 2. Result comparison using the different methods. (Color figure online)

The detailed view warped by color information in Fig. 3. The noise across the edges is almost removed, while the depth measurements are severely interfered in the original 3D view. The point cloud contains more than 300,000 points, although some points with invalid depth are eliminated. Due to the average filtering time 47 ms with GPU support, the capturing rate of the device is set to 20 fps. Then the resultant colored depth map is transmitted to a server for 3D reconstruction using SURF (Speeded Up Robust Feature) feature descriptor.

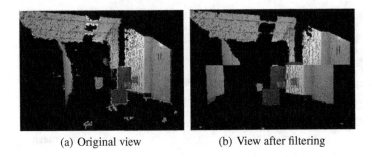

(a) Original view   (b) View after filtering

Fig. 3. Comparison of the 3D view. (Color figure online)

3.2 Quantitative Accuracy

To evaluate the performance of our filter, Cones scene from Middlebury stereo data set and video-plus-depth sequence "Interview" from ATTEST are considered. In order to illustrate the effectiveness of the confidence estimation and quantitative accuracy, confidence based noise is superimposed on the original depth image, which is considered as the ground truth. Then the filtered depth map will be compared with the ground truth. Supposed the confidence measurement C depends on the distance from the position of current pixel p to the image center c, and the noise n is subject to Gaussian distribution with mean 0, the noise model is defined as follows.

$$C = 1 - \frac{\|p - c\|}{w}$$

$$n \sim N(0, 100 \cdot (1 - C))$$

(7)

where w is a factor to balance the influence of the additional noise. Figure 4(e) shows the depth image with Gaussian noise, where the image distortion only exists across the edge, as shown in Fig. 4(d).

To evaluate the numerical accuracy, peak signal-to-noise ratio (PSNR) is employed. Given two gray-scale images I, I', usually the ratio is defined using the mean squared error (MSE).

$$MSE = \frac{1}{hw} \sum_{i=0}^{h-1} \sum_{j=0}^{w-1} \|I(i,j) - I'(i,j)\|^2$$

$$PSNR = 10 \cdot log_{10}(\frac{MAX^2}{MSE})$$

(8)

where h and w is the height and width of input image, and MAX is the maximum of the pixel value, which means $MAX = 255$, if both input are gray images, where the pixel is indicated by 8 bits. The PSNR of the depth with noise and original depth image is 24.16 dB. To speed up the implementation, only N' intensity values are selected to perform the proposed algorithm. Figure 5 shows the result of the PSNR accuracy using different number of intensity levels. When

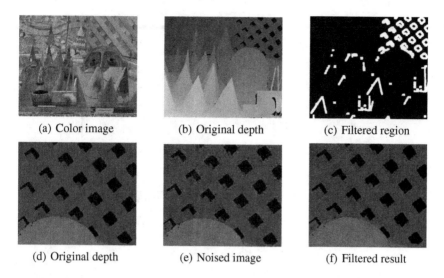

(a) Color image   (b) Original depth   (c) Filtered region

(d) Original depth   (e) Noised image   (f) Filtered result

Fig. 4. Algorithm evaluation using the cones scene from the Middlebury data set. (Color figure online)

8 levels are used, the noise level is reduced greatly, and an acceptable PSNR value is achieved, although there exists visible difference. Figure 4(f) shows the filtered result using 8 intensity levels. Note that as the level number increases, we have to compromise the computational time.

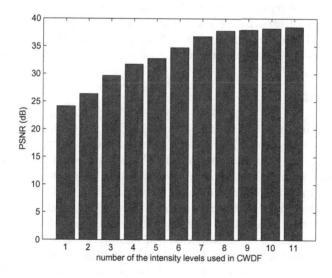

Fig. 5. PSNR accuracy using different number of intensity values.

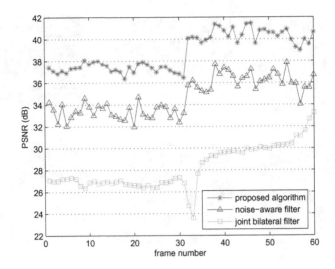

Fig. 6. PSNR accuracy with respect to different methods.

Moreover, video-plus-depth sequence "Interview" from ATTEST is used with the yuv420 video format for comprehensive test. The sequence is decoded based on the leading audio/video codec library in FFmpeg. Every frame is in the size of 720 by 576 for both color and depth sequence. The depth is indicated by 8 bits, the brighter, the closer. The first 60 frames are chosen to implement our method, joint bilateral filter and noise-aware filter respectively. Similarly, noise superimposition is repeated based on the same noise model above. Then the PSNR accuracy of each frame is calculated, as shown in Fig. 6, the proposed method has the best performance with 39 dB on average. Due to the resolution gets bigger, the GPU implementation is prolonged to the averaging 55 ms.

4   Conclusion

The accurate 3D reconstruction is required for the environmental perception and rehabilitation process. This paper presented a confidence estimation based depth filter to recover the edges around the object in 3D reconstruction of a scene with color and depth images. The proposed method takes into account both the inherent depth nature of real-time depth data and the color information from video camera to reconstruct a multi-lateral filter. Compared with the existing joint bilateral filter and noise-aware filter, the experimental results show that the proposed method obtains more detailed information around the edge of an object and reduces the geometric distortions. Using the parallel data processing based on GPU implementation, the real-time high-quality 3D reconstruction is achieved.

In future, the adaptive parameter selection for the kernel size in the range term will be achieved, while these parameters are set manually according to the different scenes in current phase. In addition, multiple depth sensors will be employed for dynamic environment. A corresponding distributed system will be set up, and more studies will concentrate on the time synchronization of multiple data streams in order to ensure the temporal stability.

References

1. Díaz Rodríguez, N., Wikström, R., Lilius, J., Cuéllar, M.P., Delgado Calvo Flores, M.: Understanding movement and interaction: an ontology for kinect-based 3D depth sensors. In: Urzaiz, G., Ochoa, S.F., Bravo, J., Chen, L.L., Oliveira, J. (eds.) UCAmI 2013. LNCS, vol. 8276, pp. 254–261. Springer, Heidelberg (2013). doi:10. 1007/978-3-319-03176-7_33
2. Suciu, D.M., Pop, B.A., Urdea, R., Mursa, B.: Non-intrusive tongue tracking and its applicability in post-stroke rehabilitation. In: Meersman, R. et al. (eds.) OTM 2014 Workshops. LNCS, vol. 8842, pp. 504–513. Springer, Heidelberg (2014)
3. Dao, T.T., Pouletaut, P., Gamet, D., Christine Ho Ba Tho, M.: Real-time rehabilitation system of systems for monitoring the biomechanical feedbacks of the musculoskeletal system. In: Nguyen, V.-H., Le, A.-C., Huynh, V.-N. (eds.) Knowledge and Systems Engineering. AISC, vol. 326, pp. 553–565. Springer, Heidelberg (2015). doi:10.1007/978-3-319-11680-8_44
4. Amzajerdian, F., Pierrottet, D., Petway, L., Hines, G., Roback, V.: Lidar systems for precision navigation and safe landing on planetary bodies. In: Proceedings of SPIE, vol. 8192 (2011)
5. Dal Mutto, C., Zanuttigh, P., Cortelazzo, G.M.: Accurate 3D Reconstruction by Stereo and ToF Data Fusion. GTTI, Brescia, Italy, June 2010
6. Lun, R., Zhao, W.: A survey of applications and human motion recognition with microsoft kinect. Int. J. Pattern Recogn. Artif. Intell. **29**(05), 1555008 (2015)
7. Kornprobst, P., Tumblin, J., do Durand, F.: Bilateral filtering: theory and applications. Found. Trends Comput. Graph. Vis. **4**(1), 1–74 (2009)
8. Kopf, J., Cohen, M.F., Lischinski, D., Uyttendaele, M.: Joint bilateral upsampling. ACM Trans. Graph. **26**, 26 (2007)
9. Chan, D., Buisman, H., Theobalt, C., Thrun, S.: A noise-aware filter for real-time depth upsampling. In: Workshop on Multi-camera and Multi-modal Sensor Fusion Algorithms and Applications(2008)
10. Huhle, B., Schairer, T., Jenke, P., Strasser, W.: Robust non-local denoising of colored depth data. In: IEEE Computer Society Conference on Computer Vision and Pattern Recognition Workshops, CVPRW 2008, pp. 1–7, June 2008
11. Kang, S.B., Szeliski, R., Chai, J.: Handling occlusions in dense multi-view stereo. In: Proceedings of the 2001 IEEE Computer Society Conference on Computer Vision and Pattern Recognition, CVPR 2001, vol. 1, pp. I-103– I-110 (2001)
12. Nvidia (2011). http://developer.nvidia.com/cuda-toolkit-40
13. Middlebury (2003). http://vision.middlebury.edu/stereo/data/scenes2003/
14. Redert, A., Op de Beeck, M., Fehn, C.,, IJsselsteijn, W., Pollefeys, M., Van Gool, L.J., Ofek, Sexton, I., Surman, P.: Attest: Advanced three-dimensional television system technologies. In: 3DPVT, pp. 313–319 (2002)

Brain-Machine Collaboration

Noisy Control About Discrete Liner Consensus Protocol

Quansheng Dou[1(✉)], Zhongzhi Shi[2], and Yuehao Pan[3]

[1] School of Computer Science and Technology,
Shandong Institute of Business and Technology, Yantai, China
douqsh@sdibt.edu.cn
[2] Key Laboratory of Intelligent Information Processing,
Institute of Computing Technology, Chinese Academy of Sciences,
Beijing, China
shizz@ics.ict.ac.cn
[3] Beijing 101 Middle School, Beijing, China
pan_yhao@yeah.net

Abstract. In recent years, more and more researchers in the control community have focused their attention on distributed coordination due to its broad applications in many fields, the consensus problem is well recognized as a fundamental problem in the cooperation control of multi-agent systems. In this paper, we discuss the noise problem of the discrete linear consensus protocol (DLCP) and point out that noise of DLCP is uncontrollable. A protocol using noise suppression function (NS-DLCP) to control the noise is put forward, the theorem about the reasonable range of the noise suppression function is vigorously proved, and sufficient conditions for noise controllable of NS-DLCP are further presented.

Keywords: Multi-agent system · Graph theory · Consensus protocol · Noisy control

1 Introduction

In recent years, more and more researchers in the control community have focused their attention on distributed coordination of multi-agent systems due to its broad applications in many fields such as sensor networks, e.g. UAV (Unmanned Air Vehicles), MRS (mobile robots systems), robotic teams.

In the cooperative control, a key problem is to design distributed protocols such that group of agents can achieve consensus through local communications. So far, numerous interesting results for consensus problem have been obtained for both discrete-time and continuous-time multi-agent system in the past decade. Reynolds systematically studied and simulated the behavior of biological group such as birds and fishes, and proposed Boidmodel [1] which still has a broad impact in the field of Swarm Intelligence. Vicsek model [2] is proposed based on statistical mechanics theory in which the movement rate of Agent on two-dimensional plane remains unchanged, and the N agents on the 2-D plane determine their motion direction according to the directions of their neighbor agents. One of the most promising tools are the linear

Z. Shi et al. (Eds.): IIP 2016, IFIP AICT 486, pp. 235–244, 2016.
DOI: 10.1007/978-3-319-48390-0_24

consensus algorithms, which are simple distributed algorithms which require only minimal computation, communication and synchronization to compute averages of local quantities that reside in each device. These algorithms have their roots in the analysis of Markov chains [3] and have been deeply studied within the computer science community for load balancing and within the linear algebra community for the asynchronous solution of linear systems [4, 5]. For linear consensus problem, Olfati-Saber et al. established a relatively complete theoretical framework based on graph theory and kinetic theory, and systematically analyzed the different types of consistency issues based on the framework [6–9]. Based on the study of Olfati-Saber, Yu et al. [10–12] discussed three necessary and sufficient conditions for the algorithm to converge to a consistent state when the states of agents has nothing to do with the data transferred, and conducted a meaningful analysis of its correctness, effectiveness and efficiency with the verification in several specific applications. For multi-Agent consensus and synchronization problems of complex networks, Li et al. in their rather deep discussion proposed multi-Agent control architecture based on higher order linear system with a series of fruitful results [13–15]. In [16] average consensus issues are discussed, with the consensus algorithm formulated as matrix factorization problem, machine learning methods are proposed to solve matrix decomposition problem.

For most of consensus results in the literature, it is usually assumed that each agent can obtain its neighbor's information precisely. Since real networks are often in uncertain communication environments, it is necessary to consider consensus problems under measurement noises. Such consensus problems have been studied, Some research [17–19] have addressed the consensus problem of multi-agents system under multiplicative measurement noises, where the noises' intensities are considered proportional to the relative states. In [20, 21] the authors studied consensus problems when there exist noisy measurements of the states of neighbors, and a stochastic approximation approach was applied to obtain mean square and almost sure convergence in models with fixed network topologies or with independent communications failures. Necessary and/or sufficient conditions for stochastic consensus of multi-agent systems were established for the case of fixed topology and time varying topologies in [22, 23]. Liu et al. studied signal delay of linear consensus protocol [24], and presented strong consensus and mean square consensus concept under the conditions of the fixed topology and the presence of noise and delay between agents, and gave theoretically necessary and sufficient conditions of strong consensus and mean square under Non_Leader_Follower and Leader_Follower modes. The distributed consensus problem for linear discrete-time multi-agent systems with delays and noises was investigated in [25] by introducing a novel technique to overcome the difficulties induced by the delays and noises. In [26], a novel kind of cluster consensus of multi-agents systems with several different subgroups was considered based on Markov chains and nonnegative matrix analysis.

In this paper, we discussed the noise problem of the linear consensus protocol, and gave a sufficient condition that ensure the noise of linear consensus protocol is controllable. The remainder of this paper is organized as follows. Some preliminaries and definitions are given in Sect. 2. in Sect. 3, we pointed out that noise of DLCP is uncontrollable. in Sect. 4, we proposed the strategy of using noise suppression function to control noise, and put forward Theorem 1 about a reasonable range of noise suppression function; Sect. 5 is devoted to show the conclusions of this paper.

2 Preliminaries

Consider n agents distributed according to a directed graph $G = (V, E)$ consisting of a set of nodes $V = \{1, 2, \ldots, n\}$ and a set of edges $E \in V \times V$. In the digraph, an edge from node i to node j is denoted as an ordered pair (i, j) where $i \neq j$ (so there is no edge between a node and itself). A path (from i_1 to i_l) consists of a sequence of nodes $i_1, i_2, \cdots, i_l, l \geq 2$, such that $(i_k, i_{k+1}) \in E$ for $k = 1 \cdots, l - 1$. We say node i is connected to node $j (i \neq j)$ if there exists a path from i to j. For convenience of exposition, the two names, agent and node, will be used alternatively. The agent A_k (resp., node k) is a neighbor of A_i (resp., node i) if $(k, i) \in E$ where $k \neq i$. Denote the neighbors of node i by $N_i = \{k | (k, i) \in E\}$. For agent A_i, we denote its state at time t by $x_i(t) \in \mathbb{R}$, where $t \in \mathbb{Z}_+, \mathbb{Z}_+ = \{0, 1, 2, \cdots\}$. For each $i \in V$, agent A_i receives information from its neighbors.

Definition 1: (Discrete Linear Consensus Protocol DLCP) The so-called linear consensus protocol is given by the following (1):

$$x_i(t+1) = x_i(t) + \sum_{j \in N_{\ominus i}(t)} \alpha_{ij}(t)\big(x_j(t) - x_i(t)\big) \quad \forall i, j \in V \tag{1}$$

Where $\alpha_{ij}(t) > 0$ is a real-valued function with variable t and $\sum\limits_{j=1}^{n} \alpha_{ij}(t) \leq 1$, it is used to characterize the extent of the impact at time t from agent j to agent i.

Definition 2: (Weighted Laplacian Matrix) The matrix $L(t) = \big[l_{ij}(t)\big]_{n \times n}$ is called weighted Laplacian matrix of graph G, where

$$l_{ij}(t) = \begin{cases} -\alpha_{ij}(t) & \text{if } j \in N_i \text{ and } i \neq j \\ \sum\limits_{j=1}^{n} \alpha_{ij}(t) & \text{if } i = j \\ 0 & \text{otherwise} \end{cases}$$

let $X(k) = [x_1(k), \ldots, x_n(k)]^T$, I_n denote an n order unit matrix, the matrix form of (1):

$$X(t+1) = A(t)X(t) \tag{2}$$

where $A(t) = I_n - L(t)$.

Suppose $r \sim N(\mu, \sigma^2)$ is a random number that satisfies the normal distribution, let var(r) represent variance σ^2 of r, if $R = [r_1, \ldots, r_n]^T$ is a random vector, then var(R) represent the covariance matrix of R, $[\text{var}(R)]_i$ denotes the variance of the ith components of R, i.e. $[\text{var}(R)]_i = \text{var}(r_i)$ is ith element of diagonal of matrix var(R).

If the received message of *Agent i* contains a mutually independent and normally distributed noise interference, (1) can be rewritten as:

$$x_i(t+1) = x_i(t) + \sum_{j \in N_i(t)} \alpha_{ij}(t)\big((x_j(t) + r_j(t)) - x_i(t)\big) \quad \forall i, j \in V \tag{3}$$

where $r_j(t) \sim N\left(0, \sigma_j^2\right)$ is a random number and satisfies normal distribution, which represents the noise carried by the status information $x_j(t)$. Let $\mathbf{R}(t) = [r_1(t), \ldots, r_n(t)]^T$, thus (3) can be rewritten in matrix form:

$$\mathbf{X}(t+1) = \mathbf{A}(t)\mathbf{X}(t) + \mathbf{W}(t)\mathbf{R}(t) \tag{4}$$

In the above equation, $\mathbf{W}(t) = -\mathbf{L}(t) - diag(-\mathbf{L}(t))$, $diag(-\mathbf{L}(t))$ is the diagonal matrix of $-\mathbf{L}(t)$, so we can further get:

$$\mathbf{X}(t+1) = \prod_{k=0}^{t} \mathbf{A}(k)\mathbf{X}(0) + \sum_{m=1}^{t}\left(\left(\prod_{j=t-m+1}^{t} \mathbf{A}(j)\right)\mathbf{W}(t-m)\mathbf{R}(t-m)\right) + \mathbf{W}(t)\mathbf{R}(t) \tag{5}$$

In order to facilitate the description, let $\left(\prod_{j=t-m+1}^{t} \mathbf{A}(j)\right) = \mathbf{I}_n$ when $m = 0$, then (5) can be simplified as:

$$\mathbf{X}(t+1) = \prod_{k=0}^{t} \mathbf{A}(k)\mathbf{X}(0) + \sum_{m=0}^{t}\left(\left(\prod_{j=t-m+1}^{t} \mathbf{A}(j)\right)\mathbf{W}(t-m)\mathbf{R}(t-m)\right)$$

let $\mathbf{R}(t-m) = \mathbf{W}(t-m)\mathbf{R}(t-m)$, $\mathbf{Y}(t) = \sum_{m=0}^{t}\left(\prod_{j=t-m+1}^{t} \mathbf{A}(j)\mathbf{R}(t-m)\right)$, $\mathbf{B}(t) = \prod_{k=0}^{t} \mathbf{A}(k)\mathbf{X}(0)$ thus we get:

$$\mathbf{X}(t+1) = \mathbf{B}(t) + \mathbf{Y}(t) \tag{6}$$

Analyzing the random part $\mathbf{Y}(t)$ of (6), we can find out that it is a linear combination of several random vectors, therefore it is also a random vector satisfying normal distribution.

Definition 3: (Noise Controllable) Assuming consensus protocol can converge to a consistent state vectors $\mathbf{X}^* = [x^*, \ldots, x^*]^T$ under the noise-free conditions, we call the consensus protocol described in (6) is noise controllable, if and only if when $t \to \infty$, $\lim_{t \to \infty} \mathbf{B}(t) = \mathbf{X}^*$ for $\forall i, j \in V$ and there is constant M which will make $\lim_{t \to \infty}[var(\mathbf{Y}(t))]_i \leq M$, $i = 1, \ldots, n$.

3   Noise Uncontrollability of Discrete Linear Consistency Protocol

For any initial state $\mathbf{X}(0)$, assuming that the consistency protocol in (2) converge to a consistent state $\mathbf{X}^*$ associated with $\mathbf{X}(0)$, under this condition, we discuss the impact of the noise on the protocol.

Lemma 1: Suppose $\mathbf{Y}(t)$ is random part of consensus protocol (6), when $t \to \infty$, $\lim_{t\to\infty}[\mathrm{var}(\mathbf{Y}(t))]_i = \infty$ for any initial state $\mathbf{X}(0)$, $i = 1, \ldots, n$.

Proof: let $y(t,m) = \prod_{j=t-m+1}^{t} A(j)R(t-m)$, then $\mathbf{Y}(t) = \sum_{m=0}^{t} (y(t,m))$ we have:

$$\mathrm{var}(y(t,m)) = A(t-m+1)var\left(\prod_{j=t-m+2}^{t} A(j)R(t-m)\right)A(t-m+1)^T$$

$$= A(t)\ldots A(t-m+1)var(R(t-m))A(t-m+1)^T\ldots A(t)^T$$

where $\mathrm{var}(R(t-m)) = W(t-m)var(R(t-m))W(t-m)^T$, $m = 1, \ldots, t$. It is known that when there is no noise, and $t \to \infty$, $\prod_{k=0}^{t} A(k)\mathbf{X}(0)$ converges, for a determined constant m, there always is:

$$\lim_{t\to\infty} var(y(t,m)) = \lim_{t\to\infty} A(t)\ldots A(t-m+1)var(R(t-m))A(t-m+1)^T\ldots A(t)^T$$

$$= \lim_{t\to\infty}(\mathbf{V}_1(m), \mathbf{V}_2(m), \ldots \mathbf{V}_n(m))A(t-m+1)^T\ldots A(t)^T$$

$$= \lim_{t\to\infty}\left(A(t)\ldots A(t-m+1)(\mathbf{V}_1(m), \mathbf{V}_2(m), \ldots \mathbf{V}_n(m))^T\right)^T = (\zeta(m))_{n\times n}$$

Where, $\mathbf{V}_i(m) = (v_i, \ldots, v_i)^T$ is a constant vector related to m, $(\zeta(m))_{n\times n}$ is a constant matrix, and $\zeta(m) > 0$. So that:

$$\lim_{t\to\infty}[\mathrm{var}(\mathbf{Y}(t))]_i = \lim_{t\to\infty}\left[var\left(\sum_{m=0}^{t}(y(t,m))\right)\right]_i = \lim_{t\to\infty}\sum_{m=0}^{t}[var(y(t,m))]_i = \infty$$

$\square$

In fact, when $t \to \infty$, $\mathbf{B}(t) = \prod_{k=0}^{t} A(k)\mathbf{X}(0)$ will eventually reach a consistent state. Similarly, for a specific constant m, $\mathrm{var}(y(t,m))$ will eventually tend to a stable constant when $t \to \infty$, and $\mathrm{var}(\mathbf{Y}(t))$ just is the infinite series accumulated by $\mathrm{var}(y(t,m))$, So it will not converge, i.e. the consensus protocol (6) is noise uncontrollable.

4 Noise Suppression Discrete Linear Consensus Protocol (NS-DLCP)

We reconstruct the state transition matrix $A(t)$, let $L_\varepsilon(t) = \varepsilon(t)L(t)$, where $\varepsilon(t):$ $\mathbb{R}_+ \to \mathbb{R}_+$ is a function whose independent variable is t, $\varepsilon(t) > 0$ and when $t \to \infty$, $\varepsilon(t) \to 0$, we call $\varepsilon(t)$ as noise suppression function, Let $A_\varepsilon(t) = I_n - \varepsilon(t)L(t)$, replace $A(t)$ in (2) with $A_\varepsilon(t)$, we get:

$$X(t+1) = A_\varepsilon(t)X(t) \tag{7}$$

Here we call (7) as Noise Suppression Consensus Protocol(NS-CP), then we rewrite (7) as the relation between $X(t+1)$ and the initial state $X(0)$, and consider the noise carried by *Agent*, then (7) is rewritten as:

$$X(t+1) = \prod_{k=0}^{t} A_\varepsilon(k)X(0) + \sum_{m=0}^{t} \left(\left(\prod_{j=t-m+1}^{t} A_\varepsilon(j) \right) W_\varepsilon(t-m)R(t-m) \right) \tag{8}$$

Similarly, let $R_\varepsilon(t-m) = W_\varepsilon(t-m)R(t-m)$, $Y_\varepsilon(t) = \sum_{m=0}^{t} \left(\prod_{j=t-m+1}^{t} A_\varepsilon(j)R_\varepsilon(t-m) \right)$,

$B_\varepsilon(t) = \prod_{k=0}^{t} A_\varepsilon(k)X(0)$ then (8) is simplified as:

$$X(t+1) = B_\varepsilon(t) + Y_\varepsilon(t) \tag{9}$$

Lemma 2: Suppose consensus protocol (2) can converge to a consistent state X^* under the noise-free conditions, if noise suppression function $\varepsilon(t)$ is the low-order infinitesimal of t^{-1}, then $\lim_{t\to\infty} B_\varepsilon(t) = X^*$.

Proof: Study formula (10), we have $X(t+1) = B_\varepsilon(t)$ in the case without noise, from the conclusion in [11] we know that $\|X(t+1) - X^*\| \le \mu_{\varepsilon 2}(t)\|X(t) - X^*\|$ for any determined t, where $\mu_{\varepsilon 2}(t)$ is the second largest eigenvalues of matrix $\frac{1}{2}\left(A_\varepsilon(t) + A_\varepsilon(t)^T\right)$, let $\lambda_2(t)$ be the second smallest eigenvalues of $\frac{1}{2}\left(L(t) + L(t)^T\right)$, obviously, $\mu_{\varepsilon 2}(t) = 1 - \varepsilon(t)\lambda_2(t)$, thus:

$$\|B_\varepsilon(t) - X^*\| \le \prod_{k=1}^{t} (1 - \varepsilon(k)\lambda_2(k))\|X(0) - X^*\| \tag{10}$$

Let λ_2^* be the smallest one in the second smallest eigenvalues of $\frac{1}{2}\left(L(t) + L(t)^T\right)$, according to the known conditions that $O(\varepsilon(t)) < O(t^{-1})$, then we can deduce that $\lim_{t\to\infty}\left(1 - \varepsilon(t)\lambda_2^*\right)^t = 0$, and because $\varepsilon(k) > 0$, thus for $\forall t$, $0 \le \prod_{k=1}^{t} (1 - \varepsilon(k)\lambda_2(k)) \le$ $\left(1 - \varepsilon(t)\lambda_2^*\right)^t$, when $t \to \infty$, from squeeze theorem we can obtain: $\lim_{t\to\infty} \prod_{k=1}^{t} (1 - \varepsilon(k)$ $\lambda_2(k)) = 0$, that means:

$$0 \leq \lim_{t \to \infty} \|\mathbf{B}_\varepsilon(t) - \mathbf{X}^*\| \leq \lim_{t \to \infty} \prod_{k=1}^{t} (1 - \varepsilon(k)\lambda_2(k)) \|\mathbf{X}(0) - \mathbf{X}^*\| = 0$$

Then we have $\lim_{t \to \infty} \|\mathbf{B}_\varepsilon(t) - \mathbf{X}^*\| = 0$, i.e. $\lim_{t \to \infty} \mathbf{B}_\varepsilon(t) = \mathbf{X}^*$.

Lemma 3: Suppose consensus protocol (2) can converge to a consistent state $\mathbf{X}^*$ under the noise-free conditions, if noise suppression function $\varepsilon(t)$ is the high-order infinitesimal of $t^{-0.5}$, then there is constant M which make $\lim_{t \to \infty} [\mathrm{var}(\mathbf{Y}_\varepsilon(t))]_i \leq M$.

Proof: let $\|\bullet\|_\infty$ to represent the row sum norm of the matrix, and investigate the row sum norm of the variance matrix of $R_\varepsilon(t - m)$, then:

$$\begin{aligned}
\|\mathrm{var}(R_\varepsilon(t-m))\|_\infty &= \left\| W_\varepsilon(t-m) var(\mathbf{R}(t-m)) W_\varepsilon(t-m)^T \right\|_\infty \\
&= \varepsilon^2(t-m) \left\| W(t-m) var(\mathbf{R}(t-m)) W(t-m)^T \right\|_\infty \\
&\leq \varepsilon^2(t-m) \|W(t-m)\|_\infty \|var(\mathbf{R}(t-m))\|_\infty \|W(t-m)^T\|_\infty \\
&\leq \varepsilon^2(t-m) \|var(\mathbf{R}(t-m))\|_\infty
\end{aligned}$$

let $y_\varepsilon(t,m) = \prod_{j=t-m+1}^{t} A_\varepsilon(j) R_\varepsilon(t-m)$, then $\mathbf{Y}_\varepsilon(t) = \sum_{m=0}^{t} (y_\varepsilon(t,m))$. Study the norm of the variance matrix of $y_\varepsilon(t,m)$, we have:

$$\begin{aligned}
\|\mathrm{var}(y_\varepsilon(t,m))\|_\infty &= \left\| A_\varepsilon(t) \ldots A_\varepsilon(t-m+1) var(R_\varepsilon(t-m)) A_\varepsilon(t-m+1)^T \ldots A_\varepsilon(t)^T \right\|_\infty \\
&\leq \|A_\varepsilon(t)\|_\infty \ldots \|A_\varepsilon(t-m+1)\|_\infty \|var(R_\varepsilon(t-m))\|_\infty \|A_\varepsilon(t-m+1)^T\|_\infty \ldots \|A_\varepsilon(t-m+1)^T\|_\infty \\
&\leq \|var(R_\varepsilon(t-m))\|_\infty \leq \varepsilon^2(t-m) \|var(\mathbf{R}(t-m))\|_\infty
\end{aligned}$$

In fact, $[\mathrm{var}(y_\varepsilon(t,m))]_i$ is exactly the ith element of the diagonal of the variance matrix $\mathrm{var}(y_\varepsilon(t,m))$, denote $\rho = \max([\mathrm{var}(y_\varepsilon(t,m))]_i)$, obviously $[\mathrm{var}(y_\varepsilon(t,m))]_i \leq \varepsilon^2(t-m)\rho$, then

$$[\mathrm{var}(\mathbf{Y}_\varepsilon(t))]_i = \sum_{m=0}^{t} [\mathrm{var}(y_\varepsilon(t,m))]_i \leq \rho(\varepsilon^2(t) + \ldots + \varepsilon^2(0)) = \rho \sum_{m=0}^{t} \varepsilon^2(m)$$

According to the condition that $O(\varepsilon(t)) > O(t^{-0.5})$, therefore series $\rho \sum_{m=0}^{t} \varepsilon^2(m)$ will converge when $t \to \infty$, let $\lim_{t \to \infty} \rho \sum_{m=0}^{t} \varepsilon^2(m) = M$, we can obtain: $\lim_{t \to \infty} [\mathrm{var}(\mathbf{Y}_\varepsilon(t))]_i \leq M$ $\qquad \square$

From Lemmas 2 and 3, it easy to get:

Theorem 1: Suppose consensus protocol (2) can converge to a consistent state $\mathbf{X}^*$ under the noise-free conditions, if order of $\varepsilon(t)$ satisfies $O(t^{-0.5}) < O(\varepsilon(t)) < O(t^{-1})$ then NS-SDLC (9) is noise controllable.

Table 1. Main conclusions of this paper

No.	$\varepsilon(t)$	$t \rightarrow \infty$, $i = 1,\ldots,n$		
		$\lim \mathbf{B}_{\varepsilon}(t)$	$\lim[\text{var}$ $(\mathbf{Y}_{\varepsilon}(t))]_i$	The final state x_i of *Agent i*
Case.1	$\varepsilon(t) = 1$	$\mathbf{X}^*$	∞	Diverge(Noise uncontrollable)
Case.2	$O(\varepsilon(t)) \leq O$ $(t^{-0.5})$	$\mathbf{X}^*$	∞	Diverge(Noise uncontrollable)
Case.3	$O(t^{-0.5}) < O$ $(\varepsilon(t)) < O(t^{-1})$	$\mathbf{X}^*$	Bounded	Normal distribution with center x^*(Noise controllable)
Case.4	$O(\varepsilon(t)) \geq O$ (t^{-1})	Unconsistent state vectors $\mathbf{X}^{\wedge}$	Bounded	Normal distribution with center $\hat{x_i}$(Noise uncontrollable)

5 Conclusion

Bases on the above theoretical results and discussion, Table 1 summarized the main conclusions of this paper.

Our main conclusions are:

I. if $\varepsilon(t) = 1$ (Equivalent to $\varepsilon(t)$ is useless) or $O(\varepsilon(t)) \leq O(t-0.5)$, the determined part $\mathbf{B}\varepsilon(t)$ of linear consensus protocol (9) can converge to consistent state vectors $\mathbf{X}^*$, but the variance of its random part $\mathbf{Y}\varepsilon(t)$ is unbounded. In this case, linear consensus protocol is noise uncontrollable.

II. When $O(\varepsilon(t)) \geq O(t-1)$, the variance of its random part $\mathbf{Y}\varepsilon(t)$ is bounded, but the determined part $\mathbf{B}\varepsilon(t)$ of linear consensus protocol can't converge to consistent state vectors $\mathbf{X}^*$, under this circumstances, linear consensus protocol is also noise uncontrollable.

III. If $O(t-0.5) < O(\varepsilon(t)) < O(t-1)$, $\mathbf{B}\varepsilon(t)$ will converge to consistent state vectors $\mathbf{X}^*$ and the variance of $\mathbf{Y}\varepsilon(t)$ is bounded, so linear consensus protocol is noise controllable. At this time, every Agent's state will be a normal distribution with center x^*.

Acknowledgments. This work was partially supported by the National Natural Science Foundation of China (Nos. 61272244, 61175053, 61173173, 61035003, 61202212), National Key Basic Research Program of China (No. 2013CB329502)

References

1. Reynolds, C.W.: Flocks, herds, and schools: a distributed behavioral model. In: Proceedings of the 14th Annual Conference on Computer Graphics and Interactive Techniques, pp. 25–34ic. ACM, New York (1987)
2. Vicsek, T., Czirok, A., Ben-Jacob, E., et al.: Novel type of phase transition in a system of self-driven particles. Phys. Rev. Lett. **75**(6), 1226–1229 (1995)

3. Seneta, E.: Non-negative Matrices and Markov Chains. John Wiley & Sons Inc, Springer (2006)
4. Frommer, A., Szyld, D.B.: On asynchronous iterations. J. Comput. Appl. Math. **123**, 201–216 (2000)
5. Strikwerda, J.C.: A probabilistic analysis of asynchronous iteration. J. Linear Algebra Appl. **349**, 125–154 (2002)
6. Olfati-Saber, R., Fax, J.A., Murray, R.M.: Consensus and cooperation in networked multi-agent systems. Proc. IEEE **95**(1), 215–233 (2007)
7. Olfati-Saber, R.: Evolutionary dynamics of behavior in social networks. In: Proceedings of the 46th IEEE Conference on Decision and Control, 12–14 December 2007 at the Hilton New Orleans Riverside in New Orleans, Louisiana USA (2007)
8. Olfati-Saber, R., Jalalkamali, P.: Coupled distributed estimation and control for mobile sensor networks. IEEE Trans. Autom. Control **57**(9), 2609–2614 (2012)
9. Olfati-Saber, R.: Flocking for multi-agent dynamic systems: algorithms and theory. IEEE Trans. Autom. Control **51**(3), 401–420 (2006)
10. Yu, C.-H., Nagpal, R.: A self-adaptive framework for modular robots in dynamic environment: theory and applications. Int. J. Robot. Res. **30**(8), 1015–1036 (2011)
11. Yu, C.-H.: Biologically-Inspired Control for Self-Adaptive Multiagent Systems. Doctoral Thesis, Harvard University (2010)
12. Yu, C.-H., Nagpal, R.: Biologically-inspired control for multi-agent self-adaptive tasks. In: Proceedings of the 24th AAAI Conference on Artifical Intelligence, 11–15 July, Atlanta, pp. 1702–1709 (2010)
13. Li, Z.K., Duan, Z.S., Chen, G.R., Huang, L.: Consensus of multi-agent systems and synchronization of complex networks: a unified viewpoint. IEEE Trans. Circuits Syst. I Regul. Pap. **57**(1), 213–224 (2010)
14. Li, Z., Duan, Z., Chen, G.: Dynamic consensus of linear multi-agent systems. Control Theory Appl. IET **5**(1), 19–28 (2011)
15. Li, Z., Liu, X., Ren, W.: L Xie Distributed tracking control for linear multiagent systems with a leader of bounded unknown input. IEEE Trans. Autom. Control **58**(2), 518–523 (2013)
16. Tran, T.M.D., Kibangou, A.Y.: Distributed design of finite-time average consensus protocols. In: 4th IFAC Workshop on Distributed Estimation and Control in Networked Systems, vol.4, pp. 227–233. Rhine Moselle Hall, Koblenz, Germany, September 2013
17. Ni, Y.H., Li, X.: Consensus seeking in multi-agent systems with multiplicative measurement noises. Syst. Control Lett. **62**(5), 430–437 (2013)
18. Djaidja, S., Wu, Q.H.: Leaderless consensus seeking in multi-agent systems under multiplicative measurement noises and switching topologies. In: Proceedings of the 33rd Chinese Control Conference, Nanjing, China, July 2014
19. Djaidja, S., Wu, Q.H.: Leader-following consensus for single-integratormulti-agent systems with multiplicative noises in directed topologies. Int. J. Syst. Sci. (2014)
20. Huang, M., Manton, J.H.: Coordination and consensus of networked agents with noisy measurements:stochastic algorithms and asymptotic behavior. SIAM J. Control Optim. **48**(1), 134–161 (2009)
21. Huang, M., Manton, J.H.: Stochastic consensus seeking with noisy and directed inter-agent communication:fixed and randomly varying topologies. Institute of Electrical and Electronics Engineers. Trans. Autom. Control **55**(1), 235–241 (2010)
22. Li, T., Zhang, J.-F.: Mean square average-consensus under measurement noises and fixed topologies: necessary and sufficient conditions. Automatica **45**(8), 1929–1936 (2009)

23. Li, T., Zhang, J.-F.: Consensus conditions of multi-agent systems with time-varying topologies and stochastic communication noises. Institute of Electrical and Electronics Engineers. Trans. Autom. Control **55**(9), 2043–2057 (2010)
24. Wen, G., Duan, Z., Li, Z., Chen, G.: Flocking of multi-agent dynamical systems with intermittent nonlinear velocity measurements. Int. J. Robust Nonlinear Control **22**(16), 1790–1805 (2012)
25. Liu, S., Xie, L., Zhang, H.: Distributed consensus for multi-agent systems with delays and noises in transmission channels. Automatica **47**(5), 920–934 (2011)
26. Chen, Y., Liu, J., Han, F., Yu, X.: On the cluster consensus of discrete-time multi-agent systems. Syst. Control Lett. **60**(7), 517–523 (2011)

Incomplete Multi-view Clustering

Hang Gao$^{(\boxtimes)}$, Yuxing Peng, and Songlei Jian

Science and Technology on Parallel and Distributed Processing Laboratory,
National University of Defense Technology, Changsha 410073, China
{gao.hang,pengyuxing,jiansonglei}@aliyun.com

Abstract. Real data often consists of multiple views (or representations). By exploiting complementary and consensus grouping information of multiple views, multi-view clustering becomes a successful practice for boosting clustering accuracy in the past decades. Recently, researchers have begun paying attention to the problem of incomplete view. Generally, they assume at least there is one complete view or only focus on two view problems. However, above assumption is often broken in real tasks. In this work, we propose an IVC algorithm for clustering with more than two incomplete views. Compared with existing works, our proposed algorithm (1) does not require any view to be complete, (2) does not limit the number of incomplete views, and (3) can handle similarity data as well as feature data. The proposed algorithm is based on the spectral graph theory and the kernel alignment principle. By aligning projections of individual views with the projection integration of all views, IVC exchanges the complementary grouping information of incomplete views. Consequently, projections of individual views are made complete and thereby resulting the consensus with accurate grouping information. Experiments on synthetic and real datasets demonstrate the effectiveness of IVC.

Keywords: Multi-view clustering · Incomplete view clustering · Spectral clustering

1 Introduction

Many datasets in real world are naturally comprised of heterogeneous views (or representations). Clustering with such type of data is commonly referred to as multi-view Clustering. With the assumption of complementary data representation and consensus decision of clusterings, multi-view clustering has the potential to dramatically increase the learning accuracy over single view clustering [1]. The main problem in multi-view clustering is how to integrate grouping information of individual views. Existing works can be roughly classified into three categories. (1) Multi-kernel learning based approach. The most representative work of this category is Multi-kernel Kmeans [2]. It first uses kernel representation for each

© IFIP International Federation for Information Processing 2016
Published by Springer International Publishing AG 2016. All Rights Reserved
Z. Shi et al. (Eds.): IIP 2016, IFIP AICT 486, pp. 245–255, 2016.
DOI: 10.1007/978-3-319-48390-0_25

view, and then it incorporates different views by seeking optimal combination of multiple kernels of different views. (2) Subspace learning based approach. It obtains a latent consensus subspace shared by multiple views and cluster the instances on the latent subspace. There are many research works in this category, including CCA-based methods [3], spectral graph based methods [4–6], matrix factorization based methods [7,8]. (3) Ensemble learning based approach. [9] takes a decision in each individual view separately and then combines all decisions of distinct views to establish a consensus decision by determining cluster agreements/disagreements.

Traditional research assumes data are complete in all views. However, in many real applications, parts of instances are not available in some views. For example, in a news story clustering task, articles are collected from different on-line news sources. Only a part of news are reported in all views. No single source includes all news. Another example is image clustering. Images are based on multiple visual and textual features. Some images have only a fraction of visual or textual feature sets.

Recently, a few attempts have been made for multi-view clustering with incomplete views. The first work to deal with incomplete view clustering was proposed in [10]. It uses one view's kernel representation as the similarity matrix and complete the incomplete view's kernel using Laplacian regularization. However, this approach requires that there exists at least one complete view containing all the instances. Shao et al. [11] relax the above constraint. They collectively complete the kernel matrices of incomplete datasets by optimizing the alignment of shared instances of the datasets. Furthermore, a clustering algorithm is proposed based on the kernel canonical correlation analysis. However, this approach focus on two view problem. It can not exploit relation among more than two views. Li et al. proposed a Partial view clustering algorithm (PVC) [12]. Based on non-negative matrix factorization (NMF), PVC works by establishing a latent subspace where the instances corresponding to the same example in different views are close to each other. PVC concentrates on two views problem. Extending PVC to more views suffers from computational problem. In most recently, Shao et al. developed an incomplete view clustering algorithm (MIC) [13]. MIC handles the situation of more than two incomplete views. With joint weighted non-negative matrix factorization, it learns a $L_{2,1}$ regularized latent subspace for multiple views. With mean value imputation initialization, MIC gives lower weights to the incomplete instances than the complete instances. During optimization, MIC pushes multiple views towards a consensus matrix iteratively. But, there are some limitations about MIC. It converges slowly and contains too many parameters, which makes it difficult to operate. Moreover, both PVC and MIC are NMF based method. Both of them inherit the limitations of NMF: (1) It cannot well deal with data with negative feature values. while in many real applications, the non-negative constraints can not be satisfied. (2) It is essentially linear, and thus cannot disclose non-linear structures hidden in data, which limits its learning ability. (3) It only deals with feature values, while in some applications we know the similarities (relationships) of instances while the detailed feature values are unavailable. Yin et al. [14] proposed a subspace learning algorithm. It utilizes a regression-like objective to learn a latent consensus representation. Besides, it explores the inter-view and intra-view relationship of the data examples by a graph regularization. However, it converges too slowly. It achieves optimal results with about one hundred iterations. This make it difficult to extend to more than two views.

In this paper, we focus on the problem of incomplete view clustering with more than two views. We propose a novel incomplete multi-view clustering (IVC) algorithm. Aiming at completing incomplete views, IVC first integrate individual views by collective spectral decomposition. Then, IVC aligns each individual with the integration respectively. In this way, complementary grouping information is shared among views and missing values of incomplete views are estimated. With estimated individual views, IVC constructs the latent consensus space. At last, clustering solution is obtained by applying the standard spectral clustering on the consensus space. As compared with previous works, the proposed algorithm has several advantages: (1) It does not require any view to be complete. (2) It does not limit the number of incomplete views. (3) It can handle similarity data (or kernel data) as well as feature data. (4) Since it has few parameters to be set, it is easy-implemented. (5) Due to the non-iterative optimization, it is efficient than most iterative algorithms such as MIC. Moreover, it shows better performance. We demonstrate it in the experiment.

The rest of this paper is organized as follows: In Sect. 2, we give a brief review of the spectral clustering and the kernel alignment principle which is our basis. Section 3 presents details of the proposed algorithm. In Sect. 4, we validate the proposed algorithm. Section 5 concludes the paper.

2   Preliminary

In this section, we give a brief review of the spectral clustering and the kernel alignment principle, which provide the necessary background and pave the way to the proposed algorithm.

2.1   Spectral Clustering

Spectral clustering is a theoretically sound and empirically successful clustering algorithm. It treats clustering as a graph partitioning problem. By making use of the spectrum graph theory, it project original data in a low-dimensional space that contains more discriminative grouping information. Algorithm 1 briefly describe the spectral clustering algorithm [15] which is the basis of our work.

The equivalent optimization formular of Algorithm 1 is Eq. 1.

$$\max_{\mathbf{U}\in\Re^{N\times M}} Trace\left(\mathbf{U}^T\mathbf{L}\mathbf{U}\right), \qquad\qquad s.t. \mathbf{U}^T\mathbf{U} = \mathbf{I} \qquad\qquad (1)$$

Algorithm 1. Normalized Spectral Clustering

Require:
 Similarity matrix: $\mathbf{S} \in \Re^{N \times N}$, Number of clusters: K;
1: Compute the symmetrical normalized lapacian matrix $\mathbf{L} = \mathbf{D}^{(-1/2)}\mathbf{S}\mathbf{D}^{(-1/2)}$.
2: Compute the first K largest eigenvectors $\mathbf{u}_1, \mathbf{u}_2, ...\mathbf{u}_K$ of $\mathbf{L}$.
3: Let $\mathbf{U} \in \Re^{N \times K}$ be the matrix containing the first k eigenvectors $\mathbf{u}_1, \mathbf{u}_2, ...\mathbf{u}_K$.
4: Normalize the rows of matrix $\mathbf{U}$ by scaling the norm to be 1.
5: For $i = 1, 2, ..., n$, let $\mathbf{y}_i \in \Re^K$ be the vector corresponding to the i-th row of $\mathbf{U}$.
6: Cluster the points $\{\mathbf{y}_i\}_{i=1,2,...,N}$ with K-means algorithm into cluster $C_1, C_2, ...C_K$.
7: **return** $\{C_i\}, i = 1, 2, ...K$;

2.2 Kernel Alignment

Kernel alignment is a measurement of similarity (or dissimilarity) between different kernels. Let $\mathbf{S}^{(1)}$ and $\mathbf{S}^{(2)}$ be two positive definite kernel matrices such that $\|\mathbf{S}^{(1)}\|_F \neq 0$ and $\|\mathbf{S}^{(2)}\|_F \neq 0$. Then, the dissimilarity between $\mathbf{S}^{(1)}$ and $\mathbf{S}^{(2)}$ is defined by Eq. (2) [16], where $\langle \mathbf{S}^{(1)}, \mathbf{S}^{(2)} \rangle_F = \sum_{i=1}^{N}\sum_{j=1}^{N} \mathbf{S}_{i,j}^{(1)}\mathbf{S}_{i,j}^{(2)}$.

$$\rho(\mathbf{S}^{(1)}, \mathbf{S}^{(2)}) = \frac{\langle \mathbf{S}^{(1)}, \mathbf{S}^{(2)} \rangle_F}{\|\mathbf{S}^{(1)}\|_F \|\mathbf{S}^{(2)}\|_F}. \tag{2}$$

3 Proposed Methods

In this section, we present the detail of the incomplete view clustering (IVC) algorithm. We first describe the IVC framework and present its objectives, and then describe the optimization procedures.

3.1 Model Description

Given V incomplete views and the similarity matrices are $\mathbf{S}^{(i)}, i = 1, 2, ..., V$. The cluster number is K. Incomplete views contain different numbers of observed values. In order to make these kernel matrices co-operable (or with the same size N), we initialize incomplete kernels by filling missing entries with the corresponding average of the column (i.e. early estimation).

 First, we exploit the discriminative grouping information of each individual view by spectral decomposition on its similarity matrix $\mathbf{S}^{(i)}, i = 1, 2, ..., V$.

$$\max_{\mathbf{U}^{(i)} \in \Re^{N \times K}} Trace\left(\mathbf{U}^{(i)^T}\mathbf{L}^{(i)}\mathbf{U}^{(i)}\right), \qquad s.t. \mathbf{U}^{(i)^T}\mathbf{U}^{(i)} = \mathbf{I} \tag{3}$$

Note that $\mathbf{U}^{(i)}$ is a recasted matrix of the original feature matrix. Each row of $\mathbf{U}^{(i)}$ is a new representation of an instance with lower dimension and more discriminative grouping information.

Next, in order to make different views consistent, we push them towards a latent consensus matrix $\mathbf{U}^*$. Because $\mathbf{U}^{(i)}$ is a projection of original feature matrix, $\mathbf{S}^{(i)} = \mathbf{U}^{(i)}\mathbf{U}^{(i)T}$ can be seen as a new kernel representation. Similarly, the latent consensus kernel can be decomposed as $\mathbf{S}^* = \mathbf{U}^*\mathbf{U}^{*T}$, where $\mathbf{U}^*$ is the latent projected matrix. Note that $\mathbf{U}^{(i)}$s are derieved by kernels with early estimation. We call $\mathbf{U}^*$ as early consensus projection.

Borrowing the idea from kernel alignment, we measure the dissimilarity between early consensus and each view by Eq. (4).

$$\rho(\mathbf{U}^*, \mathbf{U}^{(i)}) = \|\frac{\mathbf{U}^*\mathbf{U}^{*T}}{\|\mathbf{U}^*\mathbf{U}^{*T}\|_F^2} - \frac{\mathbf{U}^{(i)}\mathbf{U}^{(i)T}}{\|\mathbf{U}^{(i)}\mathbf{U}^{(i)T}\|_F^2}\|_F^2 \tag{4}$$

Minimizing the sum of dissimilarities between early consensus and all individuals, we get objective function (5), where λ_i is the tradeoff between different views and expresses the importance of view i in clustering.

$$\max_{\mathbf{U}^* \in \Re^{N \times K}} \sum_i \lambda_i \rho(\mathbf{U}^*, \mathbf{U}^{(i)}), \tag{5}$$

Since that $\|\mathbf{U}^{(i)}\mathbf{U}^{(i)T}\|_F^2 = K$, $\|\mathbf{U}^*\mathbf{U}^{*T}\|_F^2 = K$, by ignoring constant factors and trace property $(trace(\mathbf{A}\mathbf{A}^T) = \|\mathbf{A}\|_F^2)$, we rewrite the objective function (5) as follows.

$$\max_{\mathbf{U}^* \in \Re^{N \times K}} \sum_i \lambda_i Trace(\mathbf{U}^{(i)}\mathbf{U}^{(i)T}\mathbf{U}^*\mathbf{U}^{*T}) \tag{6}$$

Now, we retransmit the early consensus back to individuals. Specifically, we reorder each individual view as $\mathbf{U}^{(i)} = \begin{bmatrix} \mathbf{U}_a^{(i)} \\ \mathbf{U}_e^{(i)} \end{bmatrix}$, where $\mathbf{U}_a^{(i)}$ is the part derived by available (observed) values, while $\mathbf{U}_e^{(i)}$ is the part derived by estimated (or missing) values. Correspondingly, we reorder $\mathbf{U}^*$ as $\begin{bmatrix} \mathbf{U}_a^* \\ \mathbf{U}_e^* \end{bmatrix}$. Then, we update each $\mathbf{U}_e^{(i)}$ by aligning $\mathbf{U}^{(i)}$ with $\mathbf{U}^{(*)}$. According to Eq. (4), we get the objective function (7).

$$\max_{\mathbf{U}_e^{(i)}} Trace(\begin{bmatrix} \mathbf{U}_a^* \\ \mathbf{U}_e^* \end{bmatrix} \begin{bmatrix} \mathbf{U}_a^* \\ \mathbf{U}_e^* \end{bmatrix}^T \begin{bmatrix} \mathbf{U}_a^{(i)} \\ \mathbf{U}_e^{(i)} \end{bmatrix} \begin{bmatrix} \mathbf{U}_a^{(i)} \\ \mathbf{U}_e^{(i)} \end{bmatrix}^T) \tag{7}$$

In this way, complementary grouping information is exchanged among incomplete individuals. With updated $\mathbf{U}^{(i)}$s, we construct the final consensus U_f^* by Eq. (6). U_f^* contains more accurate grouping information than U^*. At last, we apply standard K-means clustering on U_f^* to get the final decision.

3.2   Model Training

In this subsection, we demonstrate how does IVC optimizes Eqs. (6) and (7).

By the cyclic property of the trace, we transform optimization problem (6) into (8), which is equivalent to a standard spectral clustering with graph laplacian $\sum_v \lambda_v \mathbf{U}^{(v)} . \mathbf{U}^{(v)^T}$. The solution of $\mathbf{U}^{(*)}$ is just the optimal consensus eigen vectors of all individual views.

$$\max_{\mathbf{U}^* \in \Re^{n \times K}} Trace(\mathbf{U}^{*T}(\sum_v \lambda_i \mathbf{U}^{(i)} \mathbf{U}^{(i)^T}) \mathbf{U}^*)$$

$$s.t.\ \mathbf{U}^{*T} \mathbf{U}^* = \mathbf{I} \tag{8}$$

Transforming and expanding Eq. (7) as Eq. (9), then, taking its derivative w.r.t. $\mathbf{U}_e^{(i)}$ and setting it to zero, we get the solution as in Eq. (10). To the ends, $\mathbf{U}_e^{(i)}$ is calculated.

$$\max_{\mathbf{U}_e^{(i)}} Trace(\begin{bmatrix} \mathbf{U}_a^* \mathbf{U}_a^{*T} & \mathbf{U}_a^* \mathbf{U}_e^{*T} \\ \mathbf{U}_e^* \mathbf{U}_a^{*T} & \mathbf{U}_e^* \mathbf{U}_e^{*T} \end{bmatrix} \begin{bmatrix} \mathbf{U}_a^{(i)} \mathbf{U}_a^{(i)^T} & \mathbf{U}_a^{(i)} \mathbf{U}_e^{(i)^T} \\ \mathbf{U}_e^{(i)} \mathbf{U}_a^{(i)^T} & \mathbf{U}_e^{(i)} \mathbf{U}_e^{(i)^T} \end{bmatrix}) \tag{9}$$

$$\mathbf{U}_e^{(i)} = -(\mathbf{U}_a^* \mathbf{U}_e^{*T} + \mathbf{U}_e^* \mathbf{U}_a^{*T} + 2\mathbf{U}_e^* \mathbf{U}_e^{*T})^{-1}$$
$$\times (\mathbf{U}_e^* \mathbf{U}_a^{*T} \mathbf{U}_a^{(i)} + \mathbf{U}_a^* \mathbf{U}_a^{*T} \mathbf{U}_a^{(i)} + \mathbf{U}_e^* \mathbf{U}_e^{*T} \mathbf{U}_a^{(i)} + \mathbf{U}_e^* \mathbf{U}_a^{*T} \mathbf{U}_a^{(i)}) \tag{10}$$

The specific procedure of IVC is summarized in Algorithm 2. IVC first initializes incomplete kernels with early estimation. Then, it projects each individual view into a more discriminative space by spectral decomposition. Next, IVC establishes the early consensus projection, and thereby updating individual projections. With these updated individual projections, IVC constructs the final consensus projection.

Algorithm 2. The Proposed algorithm

Require:
   Similarity matrices: $\mathbf{S}^{(i)}, i = 1, 2, ...V$; Number of clusters: K;
1: Initialize all individual $\mathbf{K}^{(i)}$s with early estimation;
2: Do spectral decomposition for all $\mathbf{S}^{(i)}$s by Equation (3);
3: Calculate consensus eigenvectors $\mathbf{U}^{(*)}$ by Equation (8);
4: Update each individual $\mathbf{U}_e^{(i)}$ by Equation (10);
5: Construct $\mathbf{U}_f^*$ with updated $\mathbf{U}^{(i)}$s by Equation (8);
6: Do K-means clustering with final consensus projection $\mathbf{U}_f^*$;
7: **return** $\{C_j\}, j = 1, 2, ...K$;

---

4   Experiment

4.1   Comparison Methods

We compare the proposed IVC with several state-of-art methods. The details of comparison methods are as follows:

IVC: IVC is the proposed approach in this work. We set equal default value for λ_i to be 1. Without prior knowledge, we treat all views equally.

MIC: Multiple incomplete view clustering [13] is one of the most recent work. It applies weighted joint non-negative matrix with $L_{2,1}$ regularization. The default co-regularization parameter set α_i and the robust parameter set β_i are all set to be 0.01 for all the views as in the original paper.

MVSpec: MVSpec is a weighted multi-kernel learning and specrtal graph theory based algorithm for multi-view clustering. It represents views through kernel matrices and optimize the intra-cluster variance function. We set the parameter p and initial weights as its original paper [2].

KADD: Integrating multiple kernels by adding them, and then running standard spectral clustering on the corresponding Laplacian. As suggested in earlier findings [17], even this seemingly simple approach often result in near optimal clustering as compared to more sophisticated approaches.

Concat: Feature concatenation is the most simple and intuitive way to integrate all the views. It concatenates features of all views and runs K-means clustering on the concatenated feature set.

We also report the best performance of complete single view. Note that the compared methods such as **KADD**, **Concat**, and **MVSpec** cannot directly deal with incomplete views. Therefore, we pre-process the incomplete views by mean imputation for these methods. We evaluate above methods by the normalized mutual information (**NMI**). Besides, we use k-means to get the clustering solution at the end, we run k-means 10 times and report the average performance.

4.2   Datasets

In this paper, we use one synthetic dataset and three real-world datasets to evaluate the comparison methods. The details of four datasets are as follows. Table 1 presents the statistics of the datasets.

Synthetic Dataset: This dataset contains three views. For each view, we sample points from a two component Gaussian mixture model as instances. There are two clusters (i.e. cluster A and cluster B). Both the features and views are correlated. Specifically, the cluster means and the covariances for the three views are listed below.

$$\begin{aligned}
\mu_A^{(1)} &= (2,2), \quad \Sigma_A^{(1)} = \begin{bmatrix} 1, 0.5 \\ 0.5, 2 \end{bmatrix}, \Sigma_B^{(1)} = \begin{bmatrix} 0.3, 0.2 \\ 0.2, 0.8 \end{bmatrix} \\
\mu_B^{(1)} &= (4,4)
\end{aligned}$$

$$\begin{aligned}
\mu_A^{(2)} &= (1,1), \quad \Sigma_A^{(2)} = \begin{bmatrix} 1.5, 0.2 \\ 0.2, 1 \end{bmatrix}, \Sigma_B^{(2)} = \begin{bmatrix} 0.3, 0.2 \\ 0.2, 0.8 \end{bmatrix} \\
\mu_B^{(2)} &= (3,3)
\end{aligned} \tag{11}$$

$$\begin{aligned}
\mu_A^{(3)} &= (1,2), \quad \Sigma_A^{(3)} = \begin{bmatrix} 1, -0.3 \\ -0.3, 1 \end{bmatrix}, \Sigma_B^{(3)} = \begin{bmatrix} 0.5, 0.2 \\ 0.2, 0.5 \end{bmatrix} \\
\mu_B^{(3)} &= (2,1)
\end{aligned}$$

Oxford Flowers Dataset (Flowers17)[1]: This dataset is composed of 17 flower categories, with 80 images for each category. Each image is described by different visual features using color, shape, and texture. χ^2 distance matrices for different flower features (color, shape, texture) are used as three different views.

Reuters Multilingual Dataset (Reuters)[2]: This dataset contains six samples of 1200 documents, balanced over the 6 labels (E21, CCAT, M11, GCAT, C15, ECAT). Each sample is made of 5 views (EN, FR, GR, IT, SP) on the same documents. The documents were initially in English, and the FR, GR, IT, and SP views corresponds to the words of their traductions respectively in French, German, Italian and Spanish.

Multi-feature digit Dataset (Mfeat) [18]: This dataset consists of features of handwritten numerals ('0'–'9') extracted from a collection of Dutch utility maps. 200 patterns per class (for a total of 2,000 patterns) have been digitized in binary images. These digits are represented in terms of the following five feature sets (files): mfeat-fou, mfeat-fac, mfeat-kar, mfeat-pix, and mfeat-zer.

Table 1. Details of the datasets

Dataset	# Instance	# Views	# Clusters
Synth	1000	3	2
Flowers	1360	3	17
Reuters	1200	5	6
Mfeat	2000	5	10

All original datasets are complete. We simulate incomplete views for them. In specific, we set incomplete ratio from 0 % to 90 % with 10 % as interval. Incomplete instances are distributed evenly in all views. Note that for each instance, it is available in at least one view.

---

[1] http://www.robots.ox.ac.uk/~vgg/data/.
[2] http://lig-membres.imag.fr/grimal/data.html.

4.3   Results

The NMIs of four datasets are plotted in Fig. 1. For synthetic data, IVC shows the best NMI. IVC, MIC and Concate preform stable even when the incomplete ratio is close to 90 %. While the NMIs of other methods drops sharply as incomplete ratio rises.

For Flowers17, all methods present the downward trends as incomplete ratio increasing. IVC shows relatively better NMI than others. MvSpec is the second best method. Note that MIC shows worst performance. The possible reason is that NMF-based method is not suitable for similarity data. (we apply MIC on kernel data of Flower17 as in original paper [13]).

As Flowers17, similar results for Reuters. IVC demonstrates slight advantage over MIC and more obvious advantage over others.

For Mfeat, in case of low incomplete ratio (i.e. when incomplete ratio is below 20 %), all methods except Concate show close NMIs. As the incomplete ratio arises, IVC shows more and more obvious superiority over others.

It can be summarized that although views are incomplete, their integration can still be more useful than single complete view. Among above multi-view methods, IVC achieves most accurate clustering for incomplete views in most cases.

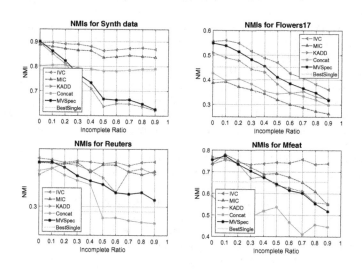

Fig. 1. NMIs

5   Conclusion

In this paper, we propose the IVC algorithm for multiple incomplete view clustering. IVC initializes incomplete views with early estimation. Based on the spectral graph theory, IVC projects original data into a new space with more discriminative grouping information. Then, individual projections are integrated. By aligning individual projections with the projection integration, estimated part of individual projections are updated to be more accurate. With those updated individual projections, final consensus is established and thereby standard K-Means is applied on. Compared with existing works, our proposed algorithm (1) does not require any view to be complete, (2) does not limit the number of incomplete views, and (3) can handle similarity data as well as feature data. Experimental results validate the effectiveness of the IVC algorithm.

Acknowledgement. This work is supported by the Major State Basic Research Development Program of China (973 Program) under the Grant No.2014CB340303, and the Natural Science Foundation under Grant No. 61402490.

References

1. Bickel, S., Scheffer, T.: Multi-view clustering. In: ICDM, vol. 4, pp. 19–26 (2004)
2. Tzortzis, G., Likas, A.: Kernel-based weighted multi-view clustering. In: 2012 IEEE 12th International Conference on Data Mining (ICDM), pp. 675–684. IEEE (2012)
3. Chaudhuri, K., Kakade, S.M., Livescu, K., Sridharan, K.: Multi-view clustering via canonical correlation analysis. In: Proceedings of the 26th Annual International Conference on Machine Learning, pp. 129–136. ACM (2009)
4. de Sa, V.R.: Spectral clustering with two views. In: ICML Workshop on Learning with Multiple Views, pp. 20–27 (2005)
5. Kumar, A., Daumé, H.: A co-training approach for multi-view spectral clustering. In: Proceedings of the 28th International Conference on Machine Learning (ICML-11), pp. 393–400 (2011)
6. Kumar, A., Rai, P., Daume, H.: Co-regularized multi-view spectral clustering. In: Advances in Neural Information Processing Systems, pp. 1413–1421 (2011)
7. Liu, J., Wang, C., Gao, J., Han, J.: Multi-view clustering via joint nonnegative matrix factorization. In: Proceedings of SDM, vol. 13, pp. 252–260. SIAM (2013)
8. Guo, Y.: Convex subspace representation learning from multi-view data. In: AAAI, vol. 1, p. 2 (2013)
9. Bruno, E., Marchand-Maillet, S.: Multiview clustering: a late fusion approach using latent models. In: Proceedings of the 32nd International ACM SIGIR Conference on Research and Development in Information Retrieval, pp. 736–737. ACM (2009)
10. Trivedi, A., Rai, P., Hal Daumé, I.I.I., DuVall, S.L.: Multiview clustering with incomplete views. In: NIPS Workshop (2010)
11. Shao, W., Shi, X., Yu, P.S.: Clustering on multiple incomplete datasets via collective kernel learning. In: 2013 IEEE 13th International Conference on Data Mining (ICDM), pp. 1181–1186. IEEE (2013)
12. Li, S.-Y., Jiang, Y., Zhou, Z.-H.: Partial multi-view clustering. In: Twenty-Eighth AAAI Conference on Artificial Intelligence (2014)
13. Shao, W., He, L., Yu, P.S.: Multiple incomplete views clustering via weighted nonnegative matrix factorization with $L_{2,1}$ regularization. In: Appice, A., Rodrigues, P.P., Santos Costa, V., Soares, C., Gama, J., Jorge, A. (eds.) ECML PKDD 2015. LNCS (LNAI), vol. 9284, pp. 318–334. Springer, Heidelberg (2015). doi:10.1007/978-3-319-23528-8_20
14. Yin, Q., Shu, W., Wang, L.: Incomplete multi-view clustering via subspace learning. In: Proceedings of the 24th ACM International on Conference on Information and Knowledge Management, pp. 383–392. ACM (2015)

15. Ng, A.Y., Jordan, M.I., Weiss, Y., et al.: On spectral clustering: analysis and an algorithm. In: Advances in Neural Information Processing Systems, vol. 2, pp. 849–856 (2002)
16. Igel, C., Glasmachers, T., Mersch, B., Pfeifer, N., Meinicke, P.: Gradient-based optimization of kernel-target alignment for sequence kernels applied to bacterial gene start detection. IEEE/ACM Trans. Comput. Biol. Bioinform. **4**(2), 216–226 (2007)
17. Cortes, C., Mohri, M., Rostamizadeh, A.: Learning non-linear combinations of kernels. In: Advances in Neural Information Processing Systems, pp. 396–404 (2009)
18. Lichman, M.: UCI machine learning repository (2013)

Brain-Machine Collaboration for Cyborg Intelligence

Zhongzhi Shi[1(✉)], Gang Ma[1,2], Shu Wang[1,2], and Jianqing Li[1,2]

[1] Key Laboratory of Intelligent Information Processing, Institute of Computing Technology,
Chinese Academy of Sciences, Beijing, 100190, China
{shizz,mag,wangs,lijq}@ics.ict.ac.cn
[2] University of Chinese Academy of Sciences, Beijing, 100049, China

Abstract. Cyborg intelligence integrates the best of both machine and biological intelligences via brain-machine integration. To make this integration effective and co-adaptive biological brain and machine should work collaboratively. Both environment awareness based collaboration and motivation based collaboration will be presented in the paper. Motivation is the cause of action and plays important roles in collaboration. The motivation leaning method and algorithm will be explored in terms of event curiosity, which is useful for sharing common interest situations.

Keywords: Cyborg intelligence · Brain-machine collaboration · Motivation driven collaboration · Motivation learning · Mind model CAM

1 Introduction

Cyborg intelligence aims to integrate AI with biological intelligence by closely and deeply connecting computer and biological beings. The term cyborg was presented by Clynes and Kline in 1960 [1], to describe a being with both organic and computing components. Combined with electronic sensing and navigation technology, a guided rat can be developed into an effective 'robot' that will possess several natural advantages over current mobile robots [2]. Supported by the National Program on Key Basic Research Project we are engaging in the research on Computational Theory and Method of Perception and Cognition of Brain-Machine Integration. The main goal is the exploration of cyborg intelligence through brain-machine integration, enhancing strengths and compensating for weaknesses by combining the biological cognition capability with the computer computational capability. To make this integration effective and co-adaptive, multi-agents should work collaboratively on environment perception, information processing, and command execution.

Collaborations occur over time as organizations interact formally and informally through repetitive sequences of negotiation, development of commitments, and execution of those commitments. Both cooperation and coordination may occur as part of the early process of collaboration, collaboration represents a longer-term integrated process. Gray describes collaboration as "a process through which parties who see different

Z. Shi et al. (Eds.): IIP 2016, IFIP AICT 486, pp. 256–266, 2016.
DOI: 10.1007/978-3-319-48390-0_26

aspects of a problem can constructively explore their differences and search for solutions that go beyond their own limited vision of what is possible" [3].

In multi-agent system no single agent owns all knowledge required for solving complex tasks. Agents work together to achieve common goals, which are beyond the capabilities of individual agent. Each agent perceives information from the environment with sensors and find out the number of cognitive tasks, even same task of having different information and the possible combination of tasks for execution for certain interval of time, pick out or select the particular combination for execution in an interval of time, and finally outputs the required effective actions to the environment.

As an internal mental model of agent, BDI model has been well recognized in philosophical and artificial intelligence area. Bratman's philosophical theory was formalized by Cohen and Levesque [4] and other researchers. A cognitive model for multi-agent collaboration should consider external perception and internal mental state of agents. Awareness is knowledge created through interaction between an agent and its environment. In multi-agent system group awareness is an understanding of the activities of others and provides a context for own activity. Group awareness can be divided into basic questions about who is collaborating, what they are doing, and where they are working. Gutwin etc. proposed a conceptual framework of workspace awareness that structures thinking about groupware interface support. They list elements for the conceptual framework [5]. Workspace awareness in a particular situation is made up of some combination of these elements.

Collaboration is goal-oriented and aided by motivation. Psychologists define motivation as an internal process that activates, guides, and maintains behavior over time. Mook defined motivation as "the cause of action" briefly [6]. Maslow proposed hierarchy of needs which was one of first unified motivation theories [7]. MicroPsi concerns modeling a motivational system to solve in the pursuit of a given set of goals, which reflects cognitive, social and physiological needs, and can account for individual variance and personality traits [8].

In this paper, a collaborative agent model for cyborg intelligence will be proposed in terms of external environment awareness and internal mental state. The agent intention is driven by motivation which is generated dynamically.

2   Cyborg Intelligence

Cyborg intelligence is dedicated to integrating AI with biological intelligence by tightly connecting machines and biological beings, for example, via brain-machine interfaces (BMIs) [9]. Figure 1 shows the physical implementation of the rat-robot navigation system [10]. In the automatic navigation of rats, five bipolar stimulating electrodes separately are implanted in medial forebrain bundle (MFB), somatosensory cortices (SI), and periaqueductal gray matter (PAG) of the rat brain. There is also a backpack fixed on the rat to receive the wireless commands.

There are two components which are necessary to implement the automatic navigation. Firstly, the communication between a computer and a rat needs to be solved. The stimulation signals are delivered by a wireless backpack stimulator which is comprised

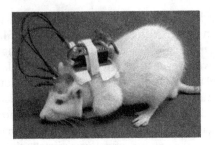

Fig. 1. Rat cyborg

of stimulating circuit, control processor and Bluetooth transceivers. The control processor receives the computer instructions through the Bluetooth transceivers. Then it sends commands to the stimulator to control the rat behaviors. By receiving commands from the machine, the rat can perform a lot of navigation tasks, e.g. walking around mazes, climbing bridges, and stopping at a special place. Secondly, a video camera device used to capture the rat movement is installed above the scenario. With the video captured by the birdeye camera, the machine can establish a map of the environment and analyze the real time kinetic state of the rat.

In brain-machine integration, each rat brain and computer can be viewed as an agent playing special role and work together for a sharing goal. The agent cognitive model is illustrated in Fig. 2, which agents deliberate the external perception and the internal mental state for decision-making. The model is represented as a 4-tuple: ⟨Awareness, Belief, Goal, Plan⟩. Awareness is described by the basic elements and relationships related to the agent's setting. Belief can be viewed as the agent's knowledge about its environment and itself. Goal represents the concrete motivations that influence an agent's behaviors. Plan is used to achieve the agent's goals. Moreover, an important module motivation-driven intention is used to drive the collaboration of cyborg intelligent system.

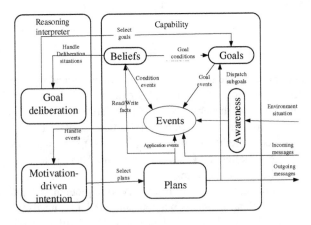

Fig. 2. Agent cognitive model

3 Environment Awareness

The environment is the complex combination of physical conditions and agents responses. Cyborg intelligent systems require bidirectional information perception between rat brain and computer. Awareness is the state or ability to perceive, to feel events, objects or sensory patterns, and cognitive reaction to a condition or event. Awareness has four basic characteristics:

- Awareness is knowledge about the state of a particular environment.
- Environments change over time, so awareness must be kept up to date.
- Agents maintain their awareness by interacting with the environment.
- Awareness establishes usually an event.

The brain computer collaborative awareness model is defined as 2-tuples: {Element, Relation}, where Element of awareness is described as follows:

(a) Who: describes the existence of agent and identity the role, answer question who is participating?
(b) What: shows agent's actions and abilities, answer question what are they doing? And what can they do? Also can show intentions to answer question what are they going to do?
(c) Where: indicates the location of agents, answer question where are they?
(d) When: shows the time point of agent behavior, answer question when can action execute?

Basic relationships contain task relationship, role relationship, operation relationship, activity relationship and cooperation relationships.

(a) Task relationships define task decomposition and composition relationships. Task involves activities with a clear and unique role attribute
(b) Role relationships describe the role relationship of agents in the multi-agent activities.
(c) Operation relationships describe the operation set of agent.
(d) Activity relationships describe activity of the role at a time.
(e) Cooperation relationships describe the interactions between agents. A partnership can be investigated through cooperation activities relevance between agents to ensure the transmission of information between different perception of the role and tasks for maintenance of the entire multi-agent perception.

Based on the integration of Marr visual theory and Gestalt whole perception theory, applying statistical learning and deep learning and other methods to analyze visual information of environment and generate high-level semantics. The convolutional generative stochastic model (CGSM) is proposed for the visual awareness [11]. The Generative Stochastic Networks (GSN) is based on learning the transition operator of a Markov chain whose stationary distribution estimates the data distribution. GSN has the capability to obtain a generative model of the data distribution without explicitly specifying a probabilistic graphical model, and allows learning deep generative model through global training via back-propagation. In order to seek a visual awareness path with a stronger robustness and a better hierarchy feature representation with gradually more global representation in the higher levels, the CGSM model can be stacked through multi convolutional generative stochastic layers. The mean-pooling is applied to CGSM so as to the downward pass operation can be successfully implemented. The layer-wise sampling like deep Boltzmann machine network will be adopted in the computational graph. CGSM has a strong robustness for noisy data, and is better to serve as a visual awareness pathway.

4   Motivation Driven Collaboration

Motivation is an internal motive force and subjective reasons, which direct drive the individual activities to achieve a certain purpose, and the psychological state initiated and maintained by individual activities. Psychologists define motivation as the process that initiates, guides, and maintains goal-oriented behaviors. All kinds of behaviors and activities of the people can't be separated from the motivation. Motivation has the following functions:

(a) Arouse the start function of the action. Personally, all the power of his actions must be through his mind, must be changed to his desire for motivation, in order to make him act up.
(b) A directing function that focuses an individual's behavior towards or away from specific goals.
(c) An organizing function that influences the combination of behavioral components into coherent, goal-oriented behavioral sequences.
(d) Strengthen function of motivation. One's experience of successes and failures on the activity, have certain influence on his activity ambition. In other words, how is the behavioral result, influencing people's motivation? Therefore the motivation plays a regulation control appearing in the form of positive or negative reinforcement role in people's behavior.

Consider the dual nature of motivation, that is implicit and explicit, the motivation process is complexity. In general, implicit motivational processes are primary and more essential than explicit motivational processes. Here we only focus on explicit motivation and hypothesize that the explicit motivational representations consist mainly of explicit goals of an agent. Explicit goals provide specific and tangible motivations for actions. Explicit goals also allow more behavioral flexibility and formation of expectancies. In cyborg intelligent system we have developed two approaches for brain computer integration, that is, needs based motivation and curiosity based motivation.

4.1   Needs Based Motivation

In 1943, humanistic psychologist Maslow put forward the demand theory of motivation. Maslow's assumption that people in need, the sequence of human motivation, from the most basic physiological and safety needs, through a series of love and respect, the complex needs of self-realization, and need level has great intuitive appeal [7]. Over the years, people have proposed a lot of theories of motivation, each theory has a certain degree of concern. These theories are very different in many ways, but they all come from a similar consideration, namely behavioral arousing, point to and keep, these three points are the core of any kind of motivation.

Bach proposed the MicroPsi architecture of motivated cognition based on situated agents [8]. MicroPsi explores the combination of a neuro-symbolic cognitive architecture with a model of autonomous, polytelic motivation. The needs of MicoPsi cognitive system fall into three groups: physiological needs, social needs and cognitive [12]. Physiological needs regulate the basic survival of the organism and reflect demands of the metabolism and physiological well-being. Social needs direct the behavior towards other individuals and groups. They are satisfied and frustrated by social signals and corresponding mental representations. Cognitive needs give rise to open-ended problem solving, skill-acquisition, exploration, play and creativity. Urges reflect various physiological, social and cognitive needs. Cognitive processes are modulated in response to the strength and urgency of the needs.

According to brain computer integration requirements, a motivation could be represented as a 3-tuples $\{N,G,I\}$, where N means needs, G is goal, I means the motivation intensity [13]. There are three type of needs in the cyborg system:

a. **Perception needs**: Acquire environment information through vision, audition, touch, taste, smell.
b. **Adaptation needs:** Adapt environment condition and optimize impaction of action.
c. **Cooperation needs:** Promise to reward a cooperation action between brain and machine.

A motivation is activated by motivational rules which structure has following format:

$$R = (P, D, \ \text{Strength}(P|D))$$

where, P indicates the conditions of rule activation; D is a set of actions for the motivation; Strength($P|D$) is a value within interval [0,1].

4.2 Curiosity Based Motivation

Curiosity based motivation is through motivation learning algorithm to build a new motivation. Agent creates internal representations of observed sensory inputs and links them to learned actions that are useful for its operation. If the result of the machine's action is not relevant to its current goal, no motivation learning takes place. This screening of what to learn is very useful since it protects machine's memory from storing unimportant observations, even though they are not predictable by the machine and may be of sufficient interest for novelty based learning. Novelty based learning still can take place in such a system, when the system is not triggered by other motivations.

Motivation learning requires a mechanism for creating abstract motivations and related goals. Once implemented, such a mechanism manages motivations, as well as selects and supervises execution of goals. Motivations emerge from interaction with the environment, and at any given stage of development, their operation is influenced by competing event and attention switching signals.

The learning process for motivations to obtain the sensory states by observing, then the sensed states are transformed mutually by the events. Where to find novelty to motivate an agent's interestingness will play an important role. Once the interestingness is stimulated, the agent's attention may be selected and focused on one aspect of the environment. Therefore, it will be necessary to define observations, events, novelty, interestingness and attention before describing the motivation learning algorithm.

Definition 1 *(Observation Functions).* Observation functions define the combinations of sensations from the sensed state that will motivate further reasoning. Observations containing fewer sensations affect an agent's attention focus by making it possible for the agent to restrict its attention to a subset of the state space. Where, a typical observation function can be given as:

$$\mathbf{O}_{S(t)} = \left\{ \left(o_{1(t)}, o_{2(t)}, \cdots, o_{L(t)}, \cdots \right) \mid o_{L(t)} = s_{L(t)} (\forall L) \right\}. \tag{1}$$

The equation defines observation function $\mathbf{O}_{S(t)}$ in which each observation focuses on every element of the sensed state at time t.

Definition 2 *(Difference Function).* A difference function Δ assigns a value to the difference between two sensations $S_{L(t)}$ and $S_{L(t')}$ in the sensed states $S_{(t)}$ and $S_{(t')}$ as follows:

$$\Delta \left(s_{L(t)}, s_{L(t')} \right) = \begin{cases} s_{L(t)}; & \text{if } \neg \exists s_{L(t')} \\ s_{L(t')}; & \text{if } \neg \exists s_{L(t)} \\ s_{L(t)} - s_{L(t')}; & \text{if } s_{L(t)} - s_{L(t')} \neq 0 \\ 0; & \text{otherwise} \end{cases} \tag{2}$$

Difference function offers the information about the change between successive sensations it calculates the magnitude of the change.

Definition 3 *(Event Function).* Event functions define which combinations of difference variables an agent recognizes as events, each of which contains only one non-zero difference variable. Event function can be defined as following formula:

$$\mathbf{E}_{S(t)} = \left\{ \mathbf{E}_{L(t)} = \left(e_{1(t)}, e_{2(t)}, \cdots, e_{L(t)}, \cdots \right) \mid e_{e(t)} \right\} \tag{3}$$

Where,

$$e_{e(t)} = \begin{cases} \Delta \left(s_{e(t)}, s_{e(t')} \right); & \text{if } e = L \\ 0; & \text{otherwise} \end{cases} \tag{4}$$

Events may be of varying length or even empty, depending on the number of sensations to change.

Definition 4 *(Novelty Detection Function)*. The novelty detection function, **N**, takes the conceptual state of the agent, $c \in C$, and compares it with memories of previous experiences, $m \in M$, constructed by long term memory to produce a novelty state, $n \in N$:

$$N : C \times M \to N. \tag{5}$$

Novelty can be detected by introspective search comparing the current conceptual state of an agent with memories of previous experiences [14].

Definition 5 *(Interestingness Function)*. The interestingness function determines a value for the interestingness of a situation, $i \in I$, basing on the novelty detected, $n \in N$:

$$I : N \to I. \tag{6}$$

Definition 6 *(Attention Selection)*. Selective attention enables you to focus on an item while mentally identifying and distinguishing the non-relevant information. In cyborg we adopt maximal interestingness strategy to select attentions to create a motivation.

The following describes the basic steps of novelty based motivation learning and goal creation algorithm in the cyborg system.

Motivation learning algorithm

(1) Observe $\mathbf{O}_{S(t)}$ from $\mathbf{S}_{(t)}$ using the observation function
(2) Subtract $\mathbf{S}_{(t)} - \mathbf{S}_{(t')}$ using the difference function
(3) Compose $\mathbf{E}_{S(t)}$ using the event function
(4) Look for $\mathbf{N}_{(t)}$ using introspective search
(5) Repeat (for each $N_i(t) \in N(t)$)
(6) Repeat (for each $I_j(t) \in I(t)$)
(7) *Attention* $= \max I_j(t)$
(8) Create a *Motivation* by *Attention*.

4.3   Motivation Execution

Motivation execution flow is shown in Fig. 3. The awareness gets information from the environment and places it into the event list. Select one event from event list and identify it. If the event is a normal event then retrieval the motivation base and select one motivation to activate the intention. If the event is a new one and never happened previously, then call motivation learning to generate a new motivation. The new motivation will activate the intention. Based on the intention plan execution will be caused and generate a series actions to accomplish the desired goal, which requires the cooperation of the reasoning machine. This means that the system will find one or more of schemes which are made in the past. It is possible to find a solution that is not the only solution when it is used to reason about an existing object. At this time, the inference engine needs to select according to its internal rules. The selection criteria need to specify before.

Different selection criteria will lead to the agent different behavioral responses at the decision-making. After choosing a good plan, the system will need to link up the goal and the plan in advance. This will make the planning a detailed understanding of the objectives, and there is sufficient information to be able to use for making planning of the goal.

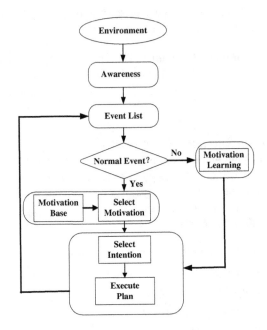

Fig. 3. Motivation execution

In cyborg system, the realization of the motivation module is through agent model ABGP. The current belief of the belief memory storage contains the agent motivation base. A desire is a goal or a desired final state. Intention is the need for the smart body to choose the current implementation of the goal. In agent, the goal is a directed acyclic graph by the sub goal composition, and realizes in step by step. According to a directed acyclic graph a sub goal is represented by a path to complete, the total goal will finish when all sub goals are completed.

4.4   Collaboration

In brain-machine integration rat brain should work with machine collaboratively. Here rat brain and machine can be abstracted as agent, so the collaboration can be viewed as joint intention [14]. Joint intention is about what the team members want to achieve. Each team member knows the intention specifically and achieves it by collaboration.

In the joint intention theory, a team is defined as "a set of agents having a shared objective and a shared mental state". The team as a whole holds joint intentions, and each team member must inform others whenever it detects the goal state change, such as goal is achieved or the goal is no longer relevant.

For the joint intention, rat agent and machine agent have three basic knowledge: first, each one should select its intention; second, each one knows its cooperator who also select the same intention; and last, each one knows they are a team. They can know each other through agent communication.

5   Conclusions

This paper described cyborg intelligence which integrates the best of both machine and biological intelligences via brain-machine integration. To make this integration effective and co-adaptive biological brain and machine should work collaboratively. Both environment awareness based collaboration and motivation based collaboration are presented in the paper. The motivation leaning algorithm is explored in terms of event curiosity, which is useful for sharing common interest situations. Under situations appear repeatedly the motivation is selected by knowledge rules.

The future of cyborg intelligence may lead towards many promising applications, such as neural intervention, medical treatment, and early diagnosis of some neurological and psychiatric disorders. Cyborg intelligence has the potential to make the bionic man reality.

The cyborg intelligence is one approach to reach the human-level intelligence. A lot of basic issues of brain-like general intelligent systems are explored in the book [15] in detail.

Acknowledgements. This work is supported by the National Program on Key Basic Research Project (973) (No. 2013CB329502), National Natural Science Foundation of China (No. 61035003), National Science and Technology Support Program (2012BA107B02).

References

1. Clynes, M.E., Kline, N.S.: Cyborgs and space. In: Astronautics, pp. 26–27, 74–76, September 1960
2. Talwar, S.K., Xu, S., Hawley, E.S., Weiss, S.A., Moxon, K.A., Chapin, J.K.: Behavioural neuroscience: rat navigation guided by remote control. Nature 417(6884), 37–38 (2002)
3. Gray, B.: Collaborating: Finding Common Ground for Multiparty Problems. Jossey-Bass, San Francisco (1989)
4. Cohen, P.R., Levesque, H.J.: Intention is choice with commitment. Artif. Intell. **42**(2–3), 213–361 (1990)
5. Gutwin, C., Greenberg, S.: The importance of awareness for team cognition in distributed collaboration. In: Salas, E., Fiore, S. (eds.) Team Cognition: Understanding the Factors that Drive Process and Performance, pp. 177–201 (2004)
6. Mook, D.G.: Motivation: The Organization of Action. W.W. Norton and Company Inc, New York (1987)

7. Maslow, A.H.: Motivation and Personality. Addison-Wesley, Boston (1954, 1970,1987)
8. Bach, J.: Principles of Synthetic Intelligence – An Architecture of Motivated Cognition. Oxford University Press, Oxford (2009)
9. Zhaohui, W., Pan, G., Carlos Príncipe, J., Cichocki, A.: Cyborg intelligence: towards bio-machine intelligent systems. IEEE Intell. Syst. **29**(6), 2–4 (2014)
10. Yu, Y., Zheng, N., Wu, Z., Zheng, X., Hua, W., Zhang, C., Pan, G.: Automatic training of ratbot for navigation. In: International Workshop on Intelligence Science, in Conjunction with IJCAI-2013, Beijing, China (2013)
11. Ma, G., Yang, X., Zhang, B., Qi, B., Shi, Z.: An environment visual awareness approach in cognitive model ABGP. In: 27th IEEE International Conference on Tools with Artificial Intelligence, pp. 744–751, November 2015
12. Bach, J.: Modeling motivation in MicroPsi 2. In: Bieger, J., Goertzel, B., Potapov, A. (eds.) AGI 2015. LNCS, vol. 9205, pp. 3–13. Springer, Heidelberg (2015)
13. Shi, Z., Zhang, J., Yue, J., Qi, B.: A motivational system for mind model CAM. In: AAAI Symposium on Integrated Cognition, Virginia, USA, pp. 79–86 (2013)
14. Shi, Z., Zhang, J., Yang, X., Ma, G., Qi, B., Yue, J.: Computational cognitive models for brain-machine collaborations. IEEE Intell. Syst. **11**(12), 24–31 (2014)
15. Shi, Z.: Mind Computation (in Chinese). Tsinghua University Press, Beijing (2015)

A Cyclic Cascaded CRFs Model for Opinion Targets Identification Based on Rules and Statistics

Hengxun Li[1(✉)], Chun Liao[2], Guangjun Hu[1], and Ning Wang[1]

[1] First Research Institute of the Ministry of Public Security of PRC,
Capital Gymnasium South Road No. 1, Haidian District, Beijing 100048, China
DerekLee1985@126.com, cityof93@qq.com, wn_1209@163.com

[2] Institute of Information Engineering, Chinese Academy of Sciences, Minzhuang Road No. 89, Haidian District, Beijing 100091, China
liaochun@iie.ac.cn

Abstract. Opinion sentences on e-commerce platform, microblog and forum contain lots of emotional information. And opinion targets identification plays an import role in huge potential commercial value mining, especially in sales decision making and development trend forecasting. Traditional CRFs-based method has achieved a pretty good result to a certain extent. However, its discovery ability of out-of-vocabulary words and optimization of the mining model are both insufficient. We propose a novel cyclic cascaded CRFs model for opinion targets identification which incorporates rule-based and statistic-based methods. The approach acquires candidate opinion targets through part-of-speech, syntactic and semantic rules, and integrates them in a cyclic cascaded CRFs model for the accurate opinion targets identification. Experimental results on COAE2014 dataset show the outperformance of this method.

Keywords: Opinion targets identification · Cyclic cascaded crfs model · Rule-based · Statistic-based

1 Introduction

With the development of the Internet, social platform has gradually integrated into people's lives, resulting in the increasing expansion of mass information. More and more opinion sentences on the Internet are generating. For the government, business or individual, the study of these opinion words is of great significance. Compared with regular grammar and news text, opinion sentences on social plat-form are more colloquial, interactive, and also contain a large number of advertisements and junk information. These bring new challenge to opinion targets identification, and how to effectively extract the useful information has become more and more important.

Sentiment analysis, also called opinion mining, is to process, induce and infer the subjective texts [1]. Sentimental elements extraction is the basis of sentiment analysis. Sentimental elements extraction is to extract the opinion elements in the sentence, including opinion words (such as "好"), opinion targets (such as "三星手机"), opinion

Z. Shi et al. (Eds.): IIP 2016, IFIP AICT 486, pp. 267–275, 2016.
DOI: 10.1007/978-3-319-48390-0_27

holder (such as "张三" in the sentence "张三认为……"). In this paper, we mainly study opinion targets identification.

Traditional CRFs-based method has achieved a pretty good result to a certain extent. However, its discovery ability of out-of-vocabulary words and optimization of the mining model are both insufficient. We propose a novel cyclic cascaded CRFs model for opinion targets identification which incorporates rule-based and statistic-based methods. The approach acquires candidate opinion targets through part-of-speech, syntactic and semantic rules, and integrates them in a cyclic cascaded CRFs model for the accurate opinion targets identification.

Existing opinion targets identification methods cannot comprehensively discovery the out-of-vocabulary words, and do not optimize the mining model. To address these shortcomings, it is intuitive to consider the combination of rule-based and statistic-based methods, and at the same time take special features of opinion sentences on social platform into consideration. In this paper, we propose a novel cyclic cascaded CRFs model for opinion targets identification which incorporates rule-based and statistic-based methods. The approach acquires candidate opinion targets through part-of-speech, syntactic and semantic rules, and integrates them in a cyclic cascaded CRFs model for the accurate opinion targets identification. In experiments on the COAE 2014 dataset we find that our method can substantially extract opinion targets more effectively under different evaluation metrics.

2   Related Work

The methods of opinion targets extraction are mainly divided into two categories: unsupervised and supervised methods. In the unsupervised methods, Hu and Liu [2] used association rules to excavate opinion targets and regarded the top-frequency words as opinion targets. Li and Zhou [3] extracted tuples like < emotional words, opinion targets > based on emotional and topic-related lexicons. Popescu and Nguyen [4] extracted properties of products with mutual information. Yao [5] used domain ontology to extract the topics and their attributes from a sentence, and summed up the subject-predicate structure rules based on syntactic analysis for opinion targets identification. Liu [6] used syntactic analysis to obtain the candidates, and then combined PMI with noun pruning algorithm to decide the final opinion targets. Besides, in the supervised methods, Zhuang [7] proposed a multi-knowledge-based approach which integrated WordNet, statistical analysis and movie knowledge. Jakob [8] modelled the task as a sequence labelling question and employed CRFs for opinion targets extraction. Wang [9] proposed a method of opinion targets identification based on CRFs, and selected morphology, dependency, relative position and semantics as features.

However, existing opinion targets extraction methods only took lexical-related features into account. Consequently, considering the specific features of Chinese microblog, we propose a new method for opinion targets extraction towards microblog using syntax and semantics in which we adopt a new approach of PDSP for domain lexicon construction and select groups of features for CRFs.

3 Candidate Opinion Targets Identification Based on Rules

The task of candidate opinion targets identification is automatically extracting the opinion targets using rule-based methods. Considering the importance of syntax and semantics in opinion targets identification, we propose a method of candidate opinion targets identification which incorporates POS, dependency structure and semantic role.

Opinion targets are usually nouns or noun phrases. Through statistics on corpus, we design six templates based on Part-of-Speech which are shown in Table 1 where n, adj, adv, aw, cmp and OTrepresents for noun, adjective, degree adverb, advocating word, comparative word and opinion target. Here we get adv and aw from Hownet, and acquire cmp from [10].

Table 1. Part-of-Speech sequence templates

Template	Example	Template	Example
n + adv + adj	屏幕/OT 很好	adj + 的+n	轻薄的机身/OT
n + adj	外观/OT 漂亮	n + cmp + n	iphone/OT 不如三星/OT
aw + n	认为蒙牛/OT	n + n	蒙牛牛奶/OT

As we all know, when we express opinions towards a product, we need some opinion words which usually have strong semantic relation with the opinion targets. Therefore, we collect opinion words from Hownet and NTUSD[1] and perform HIT-LTP[2] for dependency parsing to discuss the relation "ATT" and "SBV" between opinion words and opinion targets,relation "COO" between already known opinion targets and unknown opinion targets.

As a necessary part of shallow semantic parsing, sematic role [11] occupies an important position in lexical and semantic analysis. People usually express opinions through opinion words in opinion sentences. And adjective and verb are two main forms of opinion words. Through investigation, we find when the opinion word are adjective, A0(agent) is opinion target. Furthermore, when the opinion word are verb, A1(patient) is opinion target.

4 CCCRFs: A Cyclic Cascaded CRFs Model for Opinion Targets Identification

In this paper, we propose a novel cyclic cascaded CRFs model for opinion targets identification which incorporates rule-based and statistic-based methods. The approach first acquires candidate opinion targets through method in 3. And then, this model adopts two-layer cascaded CRFs. In the first layer, we select opinion words and manual features for opinion words identification. And in the second layer, the outputs of the first layer are added as input and we select opinion words, candidate opinion targets and manual features for opinion targets identification. In each iteration, we choose sentences whose

---

[1] http://www.datatang.com/data/11837.
[2] http://www.ltp-cloud.com/.

confidence value is larger than C as training data. And the remaining sentences are regarded as testing data.

4.1 Cascaded CRFs Model

CRFs (Conditional Random Fields, CRFs) is proposed by Lafferty [12] in 2001. Its chain structure is shown in Fig. 1.

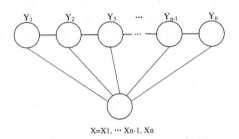

Fig. 1. CRFs model

Given a set of input random observed variables, this conditional probability distribution model can generate another set of implicit output random variables by training the model. CRFs are often used for sequence labelling tasks, such as part-of-speech tagging, named entity recognition and so on.

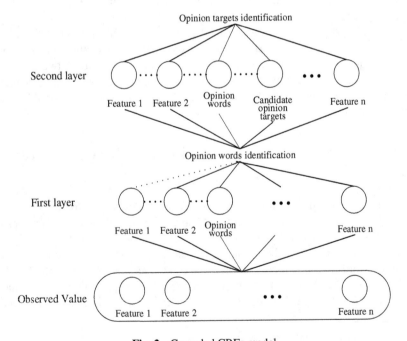

Fig. 2. Cascaded CRFs model

However, in the actual labelling process, we find it exists nesting phenomenon. For example, in Chinese named entity recognition, other named entities elements, such as names, places will be included in organization name. And this situation has led to incorrect identification. To solve this problem, you need to refine these tasks, step by step and gradually completed. In this paper, we adopt CCRFs(Cascaded Conditional Random Fields) as shown in Fig. 2 to solve the above problems.

CCRFs reduced the coupling relationship between different layers of the model. Each layer of the model can be built independently, and each sub task can be done independently without interfering with each other. The complexity has linear relationship with the number of the model layers. Before the high-level model, some of the necessary pretreatment can be carried out in the output of the underlying model and filter some errors. Consequently, CCRFs can avoid error propagation and diffusion and further improve the performance.

4.2 Cyclic Model

The selection of training data is always a focus of machine learning methods. In order to improve the recall rate of this method, we add cyclic method under CCRFs through screening each experimental results. Figure 3 is the flowchart of Cyclic Cascaded CRFs Model. If the confidence C is greater than threshold, we add them into training corpus circularly. If not, we treat them as testing data. Then, this model loops as this until the iteration number reaching a certain value N. In cyclic model, for every opinion targets identification results, its confidence degrees C, can be calculated as follows.

$$C = C_1 \times C_2 \tag{1}$$

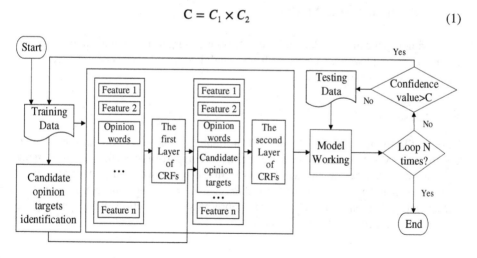

Fig. 3. Cyclic cascaded CRFs model

C_1 is the confidence value of first layer in CCRF, and C_2 is the confidence value of second layer in CCRFs. These can be acquired from CRFs tool.

We selected sentences whose confidence value larger than M into training set, and the remainder continued as the testing data. Then we re-trained the model, and extract the results for N times iteration to get the final results.

4.3   Feature Selection

In feature selection, we refer to the features which are employed by Jakob [8] and Lu [13] in English and meanwhile put forward some new features based on the specific grammar of Chinese. Generally, we think opinion targets extraction is primarily related with four kinds of features which are named as lexical features, dependency features, relative position features and semantic features.

As words with the same Part-of-Speech usually appear around the opinion targets, we select the current word itself and the POS of current word as lexical features. Dependency parsing reflects the semantic dependency relations between core word and its subsidiaries words [14]. Consequently, we select whether the dependency between current word and core word exists, the dependency type, parent word and the POS of parent word as the dependency features. As we all know, since words which appear around emotional words are more likely to be opinion targets, we determine the boolean value by judging whether the distance between current word and emotional word is less than 5. Considering there is a strong relationship between the sematic roles and POS of emotional words, we select the sematic role name of current word and POS of emotional word in this sentence for CRFs.

5   Experiments and Analysis

In experiments, we firstly obtained candidate opinion targets through method in Sect. 3, and then we employed CCCRFs with the candidate opinion targets and features in Sect. 4 together to extract opinion targets.

5.1   Dataset

Through filtration, we finally obtain 5,000 normalized sentences with opinion orientation from COAE2014. And we perform segmentation, part-of-speech, syntactic parsing and semantic role labelling through LTP [15]. In this paper, we conduct experiments on such a dataset and assess it with traditional Precision, Recall and F-measure under strict and lenient evaluations which respectively represents the extraction result is exactly the same or overlapped with the labelled one.

5.2   Parameter Selection

In this section, we make comparing experiment on parameter selection of cyclic cascaded CRFs model. And the experiment results between loop number N and confidence value C in cyclic cascaded CRFs model are in Fig. 4.

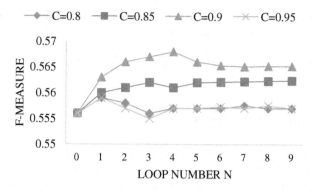

Fig. 4. Comparing experiments between loop number and confidence value in cyclic cascaded CRFs model

From this figure we find that, it performs best when the confidence value C = 0.9 and loop number N = 4. We also find the curve is slowly tend to be stable under four confidence value C with the increase of loop number N. As the confidence value C increases, opinion sentences have been added to the training corpus completely and the training corpus is no longer increased, so the model tends to a stable state.

Consequently, this experiment not only demonstrated the effectiveness of cyclic cascaded CRFs model, but also revealed the importance of parameter selection to opinion targets identification.

5.3   Comparing Results with Different Methods of Opinion Targets Identification

In this section, we compare different methods of opinion targets identification, rule-based method, CRFs-based method, cascaded CRFs-based method and cyclic cascaded CRFs-based method. To verify the effectiveness of cyclic cascaded CRFs model for opinion targets identification, we conduct experiments under different methods of opinion targets identification and obtain the experimental results as shown in Table 2. And the rule, CRFs, CCRFs and CCCRFs respectively represent for rule-based method, CRFs-based method, cascaded CRFs-based method and cyclic cascaded CRFs-based method.

Table 2.   Results of opinion targets identification with different methods

Method	Strict evaluation			Lenient evaluation		
	Precision	Recall	F-measure	Precision	Recall	F-measure
Rule	0.5870	0.3206	0.4147	0.6025	0.3971	0.4787
CRFs	0.6780	0.4325	0.5281	0.7115	0.4600	0.5490
CCRFs	0.6985	0.4595	0.5403	0.7674	0.4935	0.5729
CCCRFs	0.7085	0.4752	0.5689	0.7803	0.5025	0.6113

It can be seen that the effect of opinion targets identification is highly improved after adopting cyclic cascaded CRFs model, which is mainly because this method not only uses candidate opinion targets identification in Sect. 3 to obtain candidate opinion targets, but also adopts machine learning method of CCCRFs to make up for the defect of rule-based method and so as to reach a higher precision, recall and F-measure. So this experiment strongly demonstrates the effectiveness and applicability of cyclic cascaded CRFs model.

6   Conclusions and Future Work

In this paper we propose a cyclic cascaded CRFs model for opinion targets identification which takes rules and statistics into consideration. We combine the candidate opinion targets extraction into a cyclic cascaded CRFs model to get the final opinion targets. The experimental results show that it performs better than other baseline approaches.

In the future work, we will take the following points into consideration:

- Considering the various expressions of Chinese microblog, we should excavate more rules and extract the kernel sentence for opinion targets extraction.
- In this paper, we perform opinion targets extraction on sentence level. We will investigate the effect of opinion targets extraction on corpus level.

References

1. Zhao, Y., Qin, B., Liu, T.: Sentiment analysis. J. Softw. **21**(8), 1834–1848 (2010)
2. Hu, M., Liu, B.: Mining and summarizing customer reviews. In: Tenth ACM SIGKDD International Conference on Knowledge Discovery and Data Mining, Seattle, Washington, USA, pp. 168–177, August 2004
3. Li, B., Zhou, L., Feng, S., et al.: A unified graph model for sentence-based opinion retrieval. In: Meeting of the Association for Computational Linguistics, pp. 1367–1375. Association for Computational Linguistics (2010)
4. Popescu, A.M., Etzioni, O.: Extracting product features and opinions from reviews. In: Natural Language Processing and Text Mining, pp. 9–28. Springer, London (2007)
5. Yao, T., Lou, D.: Research on semantic orientation analysis for topics in chinese sentences. J. Chinese Inform. Process. **21**(5), 73–79 (2007)
6. Liu, H., Zhao, Y., Qin, B., Liu, T.: Comment target extraction and sentiment classification. J. Chinese Inform. Process. **24**(1), 84–88 (2010)
7. Zhuang, L., Jing, F., Zhu, X.Y.: Movie review mining and summarization. In: Proceedings of the 15th ACM International Conference on Information and Knowledge Management, pp. 43–50, November 2006
8. Jakob, N., Gurevych, I.: Extracting opinion targets in a single-and cross-domain set-ting with conditional random fields. In: Proceedings of the 2010 Conference on Empirical Methods in Natural Language Processing, pp. 1035–1045, October 2015
9. Wang, R., Ju, J., Li, S., Zhou, G.: Feature engineering for CRFs based opinion target extraction. J. Chinese Inform. Process. **26**(2), 56–61 (2012)
10. Zhang, C., Feng, C., Liu, Q., Shi, C., Huang, H., Zhou, H.: Chinese Comparative Sentence Identification Based on Multi-feature Fusion. J. Chinese Inform. Process. **27**(6), 110–116 (2013)
11. Hacioglu, K.: Semantic role labelling using dependency trees. Computational Linguistics, p. 1273, August 2004

12. Lafferty, J.D., McCallum, A., Pereira, F.C.N.: Conditional Random Fields: Probabilistic Models For Segmenting and Labelling Sequence Data, pp. 282–289 (2001)
13. Lu, B.: Identifying opinion holders and targets with dependency parser in Chinese news texts. In: Proceedings of the NAACL HLT 2010 Student Research Workshop, pp. 46–51, June 2010
14. Li, X., Roth, D.: Learning question classifiers. In: Proceedings of the 19th International Conference on Computational Linguistics, pp. 1–7, August 2002
15. Che, W., Li, Z., Liu, T.: Ltp: a Chinese language technology platform. In: Proceedings of the 23rd International Conference on Computational Linguistics: Demonstrations, pp. 13–16, August 2010

Author Index

Printed in the United States
By Bookmasters